Seed Production *of* Field Crops

Seed Production *of* Field Crops

Brijesh Tiwari

Oxford Book Company
Jaipur ▪ New Delhi

ISBN : 9789380179407

Published in India, 2026

Price : ₹1775.00

Oxford Book Company
Regd. Off. "Bony Residency", Gate No. 2
Opp. Tilak Public School, Vishweswariya Nagar,
Gopalpura Road, Jaipur - 302018, India
Ph: 0141 - 2761280, Telefax: 2761381
email - oxfordbookjpr@gmail.com
website - oxfordbookcompany.com

Delhi Office: 102, 1st Floor, Satyam House,
4327/3, Ansari Road, Darya Ganj,
New Delhi - 110002, India
Ph: 011-45652440

Distributed By
KEYA INTERNATIONAL PRESS LLP
E-mail: keyainternationalpressllp@gmail.com

Printed and bound in India by Thomson Press Ltd., New Delhi

Contents

Preface

A seed, referred to as a kernel in some plants, is a small embryonic plant enclosed in a covering called the seed coat, usually with some stored food. It is the product of the ripened ovule of gymnosperm and angiosperm plants which occurs after fertilisation and some growth within the mother plant. The formation of the seed completes the process of reproduction in seed plants, with the embryo developed from the zygote and the seed coat from the integuments of the ovule. Seed production in natural plant populations vary widely from year-to-year in response to weather variables, insects and diseases, and internal cycles within the plants themselves.

Many seeds are edible and the majority of human calories comes from seeds, especially from cereals, legumes and nuts. Seeds also provide most cooking oils, many beverages and spices and some important food additives. India has a very well organised seed industry both in public and private sectors and this can be effectively utilised for hybrid rice seed production. Realising fully well, the potential of seed industry, enough encouragement is given to the seed sectors to take up hybrid seed production. Now the seed industry has fully geared to meet the seed requirement and what it needs is the demand which has to be created by well organised transfer of technology activities.

The present book provides an updated introduction to seed production of field crops. It presents a straightforward exploration of the seed as a biological system, from its origin to its ultimate determination, and examines the fundamentals of how seeds are produced, conditioned, evaluated, and distributed in modern agriculture. This book contains chapters on seed quality and seed treatment which reflects the advances made in new physiological and molecular biology techniques. This book gives advanced undergraduate students the current state of the art and future trends in seed technology. Newcomers to the field will also benefit from the volume's straightforward style and in-depth treatment of the subject.

Brijesh Tiwari

1

Introduction to Seeds

A seed, referred to as a kernel in some plants, is a small embryonic plant enclosed in a covering called the seed coat, usually with some stored food. It is the product of the ripened ovule of gymnosperm and angiosperm plants which occurs after fertilization and some growth within the mother plant. The formation of the seed completes the process of reproduction in seed plants (started with the development of flowers and pollination), with the embryo developed from the zygote and the seed coat from the integuments of the ovule.

Seeds have been an important development in the reproduction and spread of flowering plants, relative to more primitive plants like mosses, ferns and liverworts, which do not have seeds and use other means to propagate themselves. This can be seen by the success of seed plants (both gymnosperms and angiosperms) in dominating biological niches on land, from forests to grasslands both in hot and cold climates.

The term seed also has a general meaning that predates the above — anything that can be sown i.e. "seed" potatoes, "seeds" of corn or sunflower "seeds". In the case of sunflower and corn "seeds", what is sown is the seed enclosed in a shell or hull, and the potato is a tuber.

Seeds are produced in several related groups of plants, and their manner of production distinguishes the angiosperms ("enclosed seeds") from the gymnosperms ("naked seeds"). Angiosperm seeds are produced in a hard or fleshy structure called a fruit that encloses the seeds, hence the name. (Some fruits have layers of both hard and fleshy material). In gymnosperms, no special structure develops to enclose the seeds, which begin their development "naked" on the bracts of cones. However, the seeds do become covered by the cone scales as they develop in some species of conifer.

Seed production in natural plant populations vary widely from year-to-year in response to weather variables, insects and diseases, and internal cycles within the plants themselves. Over a 20-year period, for example, forests composed of loblolly pine and shortleaf pine produced from 0 to nearly 5 million sound pine seeds per hectare. Over this period, there were six bumper seeds, five poor seeds crops, and nine good seed crops, when evaluated in regard to producing adequate seedlings for natural forest reproduction.

There are a number of modifications to seeds by different groups of plants. One example is that of the so-called stone fruits (such as the peach), where a hardened fruit layer (the endocarp) surrounds the actual seed and is fused to it.

Many structures commonly referred to as "seeds" are actually dry fruits. Sunflower seeds are sold commercially while still enclosed within the hard wall of the fruit, which must be split open to reach the seed.

STRUCTURE OF A SEED

A typical seed includes three basic parts: (1) an embryo, (2) a supply of nutrients for the embryo, and (3) a seed coat.

The embryo is an immature plant from which a new plant will grow under proper conditions. The embryo has one cotyledon or seed leaf in monocotyledons, two cotyledons in almost all dicotyledons and two or more in gymnosperms. The radicle is the embryonic root. The plumule is the embryonic shoot. The embryonic stem above the point of attachment of the cotyledon(s) is the epicotyl. The embryonic stem below the point of attachment is the hypocotyl.

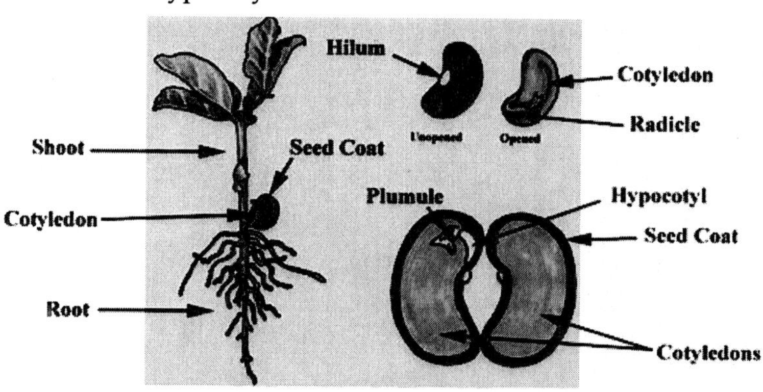

Figure 1. Seed: Internal and external structure

Within the seed, there usually is a store of nutrients for the seedling that will grow from the embryo. The form of the stored nutrition varies depending on the kind of plant. In angiosperms, the stored food begins as a tissue called the endosperm, which is derived from the parent plant via double fertilization. The usually triploid endosperm is rich in oil or starch and protein. In gymnosperms, such as conifers, the food storage tissue is part of the female gametophyte, a haploid tissue. In some species, the embryo is embedded in the endosperm or female gametophyte, which the seedling will use upon germination. In others, the endosperm is absorbed by the embryo as the latter grows within the developing seed, and the cotyledons of the embryo become filled with this stored food. At maturity, seeds of these species have no endosperm and are termed exalbuminous seeds. Some exalbuminous seeds are bean, pea, oak, walnut, squash, sunflower, and radish. Seeds with an endosperm at maturity are termed albuminous seeds. Most monocots (e.g. grasses and palms) and many dicots (e.g. brazil nut and castor bean) have albuminous seeds. All gymnosperm seeds are albuminous.

The seed coat (or testa) develops from the tissue, the integument, originally surrounding the ovule. The seed coat in the mature seed can be a paper-thin layer (e.g. peanut) or something more substantial (e.g. thick and hard in honey locust and coconut). The seed coat helps protect the embryo from mechanical injury and from drying out.

In addition to the three basic seed parts, some seeds have an appendage on the seed coat such an aril (as in yew and nutmeg) or an elaiosome (as in Corydalis) or hairs (as in cotton). There may also be a scar on the seed coat, called the hilum; it is where the seed was attached to the ovary wall by the funiculus.

SEED DEVELOPMENT

The seed, which is an embryo with two points of growth (one of which forms the stems the other the roots) is enclosed in a seed coat with some food reserves. Angiosperm seeds consist of three genetically distinct constituents: (1) the embryo formed from the zygote, (2) the endosperm, which is normally triploid, (3) the seed coat from tissue derived from the maternal tissue of the ovule. In angiosperms, the process of seed development begins with double fertilization and involves the fusion of the egg and sperm nuclei into a zygote. The second part of this process is the fusion of the polar nuclei with a second sperm cell nucleus, thus forming a primary endosperm. Right after fertilization, the zygote is mostly inactive but the primary endosperm divides rapidly to form the endosperm tissue. This tissue becomes the food that the young plant will consume until the roots have developed after germination or it develops into a hard seed coat. The seed coat forms from the two integuments or outer layers of cells of the ovule, which derive from tissue from the mother plant, the inner integument forms the tegmen and the outer forms the testa. When the seed coat forms from only one layer it is also called the testa, though not all such testa are homologous from one species to the next.

In gymnosperms, the two sperm cells transferred from the pollen do not develop seed by double fertilization but one sperm nucleus unites with the egg nucleus and the other sperm is not used. Sometimes each sperm fertilizes an egg cell and one zygote is then aborted or absorbed during early development. The seed is composed of the embryo (the result of fertilization) and tissue from the mother plant, which also form a cone around the seed in coniferous plants like Pine and Spruce. .

The ovules after fertilization develop into the seeds; the main parts of the ovule are the funicle; which attaches the ovule to the placenta, the nucellus; the main region of the ovule were the embryo sac develops, the micropyle; A small pore or opening in the ovule where the pollen tube usually enters during the process of fertilization, and the chalaza; the base of the ovule opposite the micropyle, where integument and nucellus are joined together.

The shape of the ovules as they develop often affects the finale shape of the seeds. Plants generally produce ovules of four shapes: the most common shape is called anatropous, with a curved shape. Orthotropous ovules are straight with all the parts of the ovule lined up in a long row producing an uncurved seed. Campylotropous ovules have a curved embryo sac often giving the seed a tight "c" shape. The last ovule shape is called amphitropous, where the ovule is partly inverted and turned back 90 degrees on its stalk or funicle.In the majority of flowering plants, the zygote's first division is transversely oriented in regards to the long axis, and this establishes the polarity of the embryo. The upper or chalazal pole becomes the main area of growth of the embryo, while the lower or micropylar pole produces the stalk-like suspensor that attaches to the micropyle. The suspensor absorbs and manufacturers nutrients from the endosperm that are utilized during the embryos growth.

The embryo is composed of different parts; the epicotyle will grow into the shoot, the radicle grows into the primary root, the hypocotyl connects the epicotyle and the radicle, the cotyledons form the seed leaves, the testa or seed coat forms the outer covering of the seed. Monocotyledonous plants like corn, have other structures; instead of the hypocotyle-epicotyle, it has a coleoptile that forms the first leaf and connects to the coleorhiza that connects to the primary root and adventitious roots form from the sides. The seeds of corn are constructed with these structures; pericarp, scutellum (single large cotyledon) that absorbs nutrients from the endosperm, endosperm, plumule, radicle, coleoptile and coleorhiza - these last two structures are sheath-like and enclose the plumule and radicle, acting as a protective covering. The testa or seed coats of both monocots and dicots are often marked with patterns and textured markings, or have wings or tufts of hair.

SEED SIZE AND SEED SET

Seeds are very diverse in size. The dust-like orchid seeds are the smallest with about one million seeds per gram, they are often embryonic seeds with immature embryos and no significant energy reserves. Orchids and a few other groups of plants are myco-heterotrophs which depend on mycorrhizal fungi for nutrition during germination and the early growth of the seedling. Some terrestrial Orchid seedlings, in fact, spend the first few years of their life deriving energy from the fungus and do not produce green leaves. At over 20 kg, the largest seed is the coco de mer. Plants that produce smaller seeds can generate many more seeds per flower, while plants with larger seeds invest more resources into those seeds and normally produce fewer seeds. Small seeds are quicker to ripen and can be dispersed sooner, so fall blooming plants often have small seeds. Many annual plants produce great quantities of smaller seeds; this helps to ensure that at least a few will end in a favorable place for growth. Herbaceous perennials and woody plants often have larger seeds, they can produce seeds over many years, and larger seeds have more energy reserves for germination and seedling growth and produce larger, more established seedlings after germination.

SEED FUNCTIONS

Seeds serve several functions for the plants that produce them. Key among these functions are nourishment of the embryo, dispersal to a new location, and dormancy during unfavorable conditions. Seeds fundamentally are a means of reproduction and most seeds are the product of sexual reproduction which produces a remixing of genetic material and phenotype variability that natural selection acts on.

Seeds protect and nourish the embryo or baby plant. Seeds usually give a seedling a faster start than a sporeling from a spore, because of the larger food reserves in the seed and the multicellularity of the enclosed embryo.

SEED DISPERSAL

Unlike animals, plants are limited in their ability to seek out favorable conditions for life and growth. As a result, plants have evolved many ways to disperse their offspring by dispersing their seeds. A seed must somehow "arrive" at a location and be there at a time favorable for germination and growth. When the fruits open and release their seeds in a regular way, it is called dehiscent, which is often distinctive for related groups of plants, these fruits include; Capsules, follicles, legumes, silicles and siliques. When fruits do not open and release their seeds in a regular fashion they are called indehiscent, which include these fruits; Achenes, caryopsis, nuts, samaras, and utricles.

Seed dispersal is seen most obviously in fruits; however many seeds aid in their own dispersal. Some kinds of seeds are dispersed while still inside a fruit or cone, which later opens or disintegrates to release the seeds. Other seeds are expelled or released from the fruit prior to dispersal. For example, milkweeds produce a fruit type, known as a follicle, that splits open along one side to release the seeds. Iris capsules split into three "valves" to release their seeds.

SEED GERMINATION

Seed germination is a process by which a seed embryo develops into a seedling and subsequently a plant. It involves the reactivation of the metabolic pathways that lead to growth and the emergence of the radicle or seed root and plumule or shoot.

Three fundamental conditions must exist before germination can occur. (1) The embryo must be alive, called seed viability. (2) Any dormancy requirements that prevent germination must be overcome. (3) The proper environmental conditions must exist for germination.

Seed viability is the ability of the embryo to germinate and is affected by a number of different conditions. Some plants do not produce seeds that have functional complete embryos or the seed may have no embryo at all, often called empty seeds. Predators and pathogens can damage or kill the seed while it is still in the fruit or after it is dispersed. Environmental conditions like flooding or heat can kill the seed before or during germination. The age of the seed affects its health and germination ability: since the seed has a living embryo, over time cells die and cannot be replaced. Some seeds can live for a long time before germination, while others can only survive for a short period after dispersal before they die. Seed vigor is a measure of the quality of seed; and involves the viability of the seed, the germination percentage, germination rate and the strength of the seedlings produced.

The germination percentage is simply the proportion of seeds that germinate from all seeds subject to the right conditions for growth. The germination rate is the length of time it takes for the seeds to germinate. Germination percentages and rates are affected by seed viability, dormancy and environmental effects that impact on the seed and seedling. In agriculture and horticulture quality seeds have high viability, measured by germination percentage plus the rate of germination. This is given as a percent of germination over a certain amount of time, 90% germination in 20 days, for example. 'Dormancy' is covered above; many plants produce seeds with varying degrees of dormancy, and different seeds from the same fruit can have different degrees of dormancy. It's possible to have seeds with no dormancy if they are dispersed right away and do not dry (if the seeds dry they go into physiological dormancy). There is great variation amongst plants and a dormant seed is still a viable seed even though the germination rate might be very low.

Environmental conditions effecting seed germination include; water, oxygen, temperature and light.

Three distinct phases of seed germination occur: water imbibition; lag phase; and radicle emergence.

In order for the seed coat to split, the embryo must imbibe (soak up water), which causes it to swell, splitting the seed coat. However, the nature of the seed coat determines how rapidly water can penetrate and subsequently initiate germination. The rate of imbibition is dependent on the permeability of the seed coat, amount of water in the environment and the area of contact the seed has to the source of water. For some seeds, imbibing too much water too quickly can kill the seed. For some seeds, once water is imbibed the germination process cannot be stopped, and if the seed dries out again it is fatal. Other species have seeds that can imbibe and lose water a few times without causing ill effects to the seed, and drying can cause secondary dormancy.

A number of different strategies are used by gardeners and horticulturists to break seed dormancy.

Scarification which allows water and gases to penetrate into the seed, include methods that physically break the hard seed coats or soften them by chemicals. Means of scarification include soaking in hot water or poking holes in the seed with a pin or rubbing them on sandpaper or cracking with a press or hammer. Soaking the seeds in solvents or acids is also effective for many seeds. Sometimes fruits are harvested while the seeds are still immature and the seed coat is not fully developed and sown right away before the seed coat become impermeable. Under natural conditions seed coats are worn down by rodents chewing on the seed, the seeds rubbing against rocks (seeds are moved by the wind or water currents), by undergoing freezing and thawing of surface water, or passing through an animal's digestive tract. In the latter case, the seed coat protects the seed from digestion, while often weakening the seed coat such that the embryo is ready to sprout when it gets deposited (along with a bit of fertilizer) far from the parent plant. Microorganisms are often effective in breaking down hard seed coats and are sometimes used by people as a treatment, the seeds are stored in a moist warm sandy medium for several months under non-sterile conditions.

Stratification also called moist-chilling is a method to break down physiological dormancy and involves the addition of moisture to the seeds so they imbibe water and then the seeds are subject to a period of moist chilling to after-ripen the embryo. Sowing outside in late summer and fall and allowing to overwinter outside under cool conditions is an effective way to stratify seeds, some seeds respond more favorably to periods of osculating temperatures which are part of the natural environment.

Leaching or the soaking in water removes chemical inhibitors in some seeds that prevent germination. Rain and melting snow naturally accomplish this task. For seeds planted in gardens, running water is best - if soaked in a container, 12 to 24 hours of soaking is sufficient. Soaking longer, especially in stagnant water that is not changed, can result in oxygen starvation and seed death. Seeds with hard seed coats can be soaked in hot water to break open the impermeable cell layers that prevent water intake.

Other methods used to assist in the germination of seeds that have dormancy include prechilling, predrying, daily alternation of temperature, light exposure, potassium nitrate, the use of plant growth regulators like gibberellins, cytokinins, ethylene, thiourea, sodium hypochlorite plus others.Some seeds germinate best after a fire, for some seeds fire cracks hard seed coats while in other seeds chemical dormancy is broken in reaction to the presence of smoke, liquid smoke is often used by gardeners to assist in the germination of these species.

COLLECTION OF SEEDS

There is a great variety of methods and equipment available for collection of fruits and the choice depends on a number of factors which are summarised as follows:

1. Relative size and numbers of the natural dispersal units and of the units which can be conveniently collected by man. In the case of 1 – 3 large seeds inside a dehiscent or indehiscent fruit (e.g. *Aesculus, Tectona*), collection can be done most easily by awaiting natural fall of seed or fruit and collecting from the ground. At the other extreme collection from the tree of fruiting heads of *Adina cordifolia* at 200 per kg is the only practicable way to collect the seeds; at 11 million per kg, it would be impossible to collect them after dispersal.
2. Characteristics of the fruit: size, number, position and distribution of fruits; resistance of peduncles to shaking, pulling, breaking or cutting; interval between ripening and opening.
3. Characteristics of the tree: diameter, shape and length of bole, bark thickness; shape of crown; size, angle, density and resistance to breakage of branches; density of foliage and depth of crown.
4. Characteristics of the stand: distribution and stocking of trees (e.g. isolated trees, open or dense stand); density of understorey and ground vegetation).
5. Characteristics of the site: slope, accessibility.

The various collection methods may be classified into the following: (a) Collection of fallen fruits or seeds from the forest floor, (b) Collection from the crowns of felled trees, (c) Collection from standing trees with access from the ground, (d) Collection from standing trees with access by climbing and (e) Collection from standing trees with other means of access.

Natural Seedfall

Collection from the forest floor of fruits which have fallen after natural ripening and abscission is common practice with a number of large-fruited genera. It is cheap and does not require as highly skilled labour as, for example, climbing; school children or casual labour may be used. Fruit size is very important as the larger the fruit, the easier it is to see and pick up by hand. Temperate genera commonly collected from the ground are *Quercus, Fagus, Castanea* and tropical genera include *Tectona, Gmelina, Triplochiton* and several genera among the dipterocarps.

The main disadvantages of collection from natural fruit shedding are the risks of collecting immature, empty or unsound seeds, of seed deterioration or premature germination if collection is delayed, and uncertainty in identifying the mother trees from which seed is collected. Seeds in the first fruits to fall naturally in the season are often of poor quality. In Thailand shedding of teak fruits starts in March, but observations have shown that the most viable fruits are shed in the latter part of the season, so collection is usually postponed until April. Clearing the forest floor of vegetation and debris, including old or prematurely fallen fruits, and/or spreading out sheeting of light canvas, calico or plastic, to catch the seed, can greatly facilitate collecting efficiency. If carefully timed, this operation will also eliminate much of the risk of collecting empty or non-viable seed.

Sound fruits should be gathered as soon as possible after they have fallen, to avoid damage or losses from insects, rodents or fungi and premature germination. This is of particular importance in the moist tropical forest. Observations have indicated that many of the seeds of the more important dipterocarps lose their viability within a few days of shedding and studies on *Shorea platyclados* in Malaysia demonstrated that seed lots collected from the ground included considerably more defective seeds than lots collected from the standing tree. Collection from the ground must, therefore, be perfectly timed with seedfall.

In the Jari region of Amazonia in Brazil, Woessner and McNabb found that collection of green or yellow fruits of *Gmelina arborea* from the forest floor gave the best results in operations for collecting about 10,000 kg of seed a year. They could be stored temporarily in sacks during transit from the field to the fruit processing depot, without serious loss of viability. Older brown or black fruits ferment and heat in the sacks and rapidly lose viability. Collecting teams are instructed to collect only the fresh green and yellow fruits. 50 kg of fruits can be collected per man in an 8 hour working day and yield about 3 kg of dried stones. Similar results have been obtained in

Malaysia where green and yellow fruits collected from the ground gave over 90 % germination, but brown fruits, on the other hand, produced only 53 %.

Seeds of some hard-coated species may remain viable on the forest floor for years, especially in temperate conditions. In Hungary seed of *Robinia pseudoacacia* is collected from the forest floor under 30-year-old stands in the Pusztavacs forest district. A special machine screens the top 10 cm of soil and yields about 770 kg of seed per hectare, which is the equivalent of the yield of approximately 10 seed years. Even in the tropics viable hard-coated seeds may be obtained by screening the soil below the mother trees. This has been done in Malaysia for *Parkia javanica* and *Intsia palembanica* of which the seeds are large enough to be picked up by hand. Where the seeds are smaller e.g. in *Albizzia falcataria,* sieving with a wire mesh may be more practical.

The identity of the mother tree is often uncertain when fruits are collected from the ground. Isolated trees present no problem in this respect (though they may be undesirable parents because of the risk of selfing), but much mixing of fruits can occur in dense monocultures with interlocking crowns. This is of no concern in collecting commercial quantities of seed, provided that the genetic quality of the stand is average or above average.

For research and breeding purposes, it is often necessary to maintain the identity of the mother tree of each seed lot. In such cases it is advisable to clear the grond of already fallen fruits and to accelerate the fall of new fruits by shaking, beating or cutting off branches, or climbing and picking fruits in the crown. A compromise solution, suitable for commercial collections in unimproved stands containing a mixture of good and bad phenotypic trees, is to collect fruits only below the better seed bearers and within half the radius of projection of their crowns.

Manual Shaking

If fruits are easily detached but natural fruit fall is insufficiently concentrated in time, fruit fall may be induced by artificial means. Trunks of small trees and low branches may be shaken directly by hand. Higher branches may be shaken by means of a long pole and hook or by a rope. This method has produced good results in *Cordia alliodora* and *Cedrela* spp., as it facilitates rapid collection of seed with good viability as soon as visual inspection shows that the fruits are mature.

Use of a rope involves an initial operation to pass the rope over the branch to be shaken, which is described by Robbins et al. The same method is used to hoist a saw or pulley into the crown. A thin line is attached to a weight, which is projected over the branch by hand or by catapult.

For higher branches the line may be attached to an arrow, which is shot from a bow, or to an iron rod shot from a calibre .22 rifle. A light nylon line such as a fishing line of 50 lbs. (23 kg) breaking strain is suitable and the weight or projectile used should be heavy enough to drop to the ground pulling the line with it over the branch.

Care must be taken to ensure that the line unreels readily without getting entangled, for example by using an appropriate fishing reel. Once the end of the line has reached the ground, the weight or arrow can be detached and a nylon cord of 3 – 4 mm attached instead; the line is drawn back over the branch, pulling the cord with it. The two ends of the loop can then be pulled together to shake the branch. The cord should be positioned towards the end of the branch where it will have the maximum shaking effect and not close to the bole where the branch is thickest.

In New Zealand a 30 kg-pull fibreglass long-bow firing arrows 0.8–0.9 m long and 40 g in weight has been used to project a line of 5.5 kg breaking strain over a branch. This is used, successively, to pull up a 20 kg line and a 180 kg line with pulley. In Canada both a .45 calibre gun firing a steel rod of 270 g weight and a .22 calibre gun firing a cylinder of 230–300 g weight and 2 ½ inch diameter have been used successfully up to heights varying from 20–50 m. Blank cartridges were used. The line attached to the projectile was 32 kg monofilament fishing line, which was used to haul up a 320 kg plastic twine carrying a branch breaking or cutting device.

Mechanical Shaking

Mechanical tree shakers were developed originally for fruit and nut orchards but since about 1965 the technique has been used for certain forest trees, particularly for the southern pines in the USA. The machines are expensive, need flat ground for efficient use, and experienced operators are essential to avoid excessive damage to the trees. Many cones are removed by a few seconds of shaking, but longer shaking breaks off pieces of tops and limbs. Tree shakers have no role in diffuse collecting operations in natural forest, but will probably continue to be used in intensively managed seed orchards or seed stands of a limited range of species.

The American Shock Wave tree shaker is mounted on a short wheel base truck chassis, equipped with an automatic transmission. It has a padded clamping device mounted on the extreme end of a 6 metre boom capable of clamping the trunk of a tree up to 90 cm diameter. A shake pattern is developed by counterrotation unbalanced weights in the shaker of varying frequency from 400 – 4000 cycles per minute.

The tree is generally clamped about 3 m above the ground and one 15-second shake is sufficient to remove about 80 percent of the cones from *P. elliottii*; but *P. taeda* and *P. echinata* are more difficult and a good operator may often remove only 25 – 30 percent of the cones after prolonged shaking. Ripe

P. elliottii cones require a force of about 2 kg for removal whereas *P. taeda* cones are detached only by a force of 20 kg or more. An unsuccessful attempt to reduce the force needed to remove the cones of *P. taeda* by the use of abscission inducing chemicals is reported by McLemore. The repeated shaking of the more difficult trees can cause bark rupture and breakage of the leading shoot.

With five mechanical shakers the Louisiana Forestry Commission harvested cones from 34,680 *P. elliottii* trees in twenty days. Three-quarters of the trees released 85 percent of their cones during shaking periods from 6 – 30 seconds. Performance in one hour exceeded that of a climber in a week. Results reported by McLemore and Chappell showed that the mechanical shaking of *Pinus elliottii* trees by trained operators did not harm future cone production, tree growth or vigour over the following four years.

Mechanical tree shakers are now widely used in the southeastern USA for harvesting seed in seed orchards of pines. In the case of species in which ripe cones can be easily detached, such as *P. elliottii* and *P. palustris,* trees are shaken in the period after cones have reached maturity but before they open, and the cones with their contained seeds are collected from the ground. In the case of species with persistent cones, such as *P. taeda* and *P. echinata,* shaking is deferred until the cones have opened and the objective is to shake out the seeds from the cones.

In the U.S.S.R. output from the VUS-2 vibrator working on *Pinus sibirica* was reported to be 10 – 20 times that obtained by manual methods. More recent experience with a tree-shaking machine developed in Soviet Central Asia has shown that 90 – 100 % of fruits of *Juglans, Malus, Prunus, Fraxinus* and *Gleditsia* can be collected with little stem damage. The machine is a grab-type vibrator mounted on a hydraulically operated arm on the 3-point linkage of a tractor. Optimum shaking time is 10 – 25 seconds and optimum vibration rate for most species 1000 cycles per minute. Italian CECMA vibrator machines, developed primarily for harvesting the olive crop, have also been used successfully for collecting cones of Mediterranean pines.

Dispersed Seed Collection

Although collection from the ground is most often used for fruits, it can also be used for seeds dispersed after the cones or fruits have opened. The seeds of southern pines in the USA, e.g. *Pinus elliottii* and *P. taeda* have a very short period between reaching maturity and being dispersed, and a number of methods have been evaluated for collecting loose seeds as they are dispersed. In addition to the use of sheeting spread on the ground, mentioned previously, they they include polypropylene netting round the crowns, funnel-shaped wooden frames covered with cloth or polyethylene and attached to a central hub surrounding the stem, and sheeting or nets raised on poles above the ground. Earlier experience was not very successful, as rarely more than 50 % of the available seed crop was recovered. If seed is borne mainly on or near the outside of the crown, much of it falls outside the spread of a single tree catching unit. If sheeting or netting must remain in place for an extended period of natural seed fall, it is liable to get damaged from weather and a proportion of the seeds are lost to birds and animals.

More recently a net retrieval system, for use in seed orchards in southern USA, has been developed jointly by the Georgia Forestry Commission and the Missoula Equipment Development Center. It shows considerable promise and is increasingly preferred to vacuum harvesters in *P. taeda* orchards. Netting used is a polypropylene plastic fabric manufactured for carpet backing. It is light, durable and available in a range of dimensions. For *Pinus taeda* a width of 16.5 ft (5m) and a weave count of 6 × 8 per square inch (approximately 2 × 3 per cm^2) is used.

If carefully handled, the expected life is 10+ years. Netting is laid in the orchard several weeks before seed fall and trees are shaken by mechanical shaker to dislodge seeds from the cones. The power take off from a wheeled tractor provides the power (less the 30 h.p.) needed for (a) Mechanical winding of the net onto a roller and (b) Mechanical separation of seed from the twigs, leaves and cones which also fall to the ground as a result of the shaking operation. Care must be taken in separating the netting from the grass blades which tend to go through it, before starting to wind in the netting.

Gathering of fruits on the ground is usually done manually but may be assisted by using a simple hand tool, such as a long-handled rake with interchangeable heads having different numbers of, and spacing between, tines. Attempts have been made to develop mechanical or vacuum sweeper methods to pick up seeds or fruits. Tests have also been made of a mechanical harvester, with the sweeping action of a rotating drum fitted with a myriad of attached rubber fingers which pick up seeds with separation from foreign debris. These types of machine are best used in conjunction with a tree shaker, which ensures that a substantial quantity of seeds are on the ground at each sweeping operation.

Mineau described the successful use in France of a vacuum sweeper type of machine, operated by compression from a tractor engine, for collection of *Fagus* fruits. The machine is compact and weighs 450 kg. A Dutch vacuum seed harvester has been used successfully for the collection of *Quercus* acorns and is reported to be a cheaper method in the Netherlands than either hand-picking or the use of a tree shaker and tarpaulins.

Animal Caches

Animals sometimes gather together cones or fruits as a food supply and these caches may be raided for the seed, but this source of seeds is confined to limited areas. Squirrel caches are an important source of coniferous seed in western North America. Squirrels usually locate their caches year after year in the same places. Typically, they are found in damp areas near springs, small creeks or marshes, on northern exposures, and in decayed wood or duff or around old fallen trees. A single cache may contain from a few cones to many bushels. Fresh cones on the ground are a sign of squirrel activity; piles of cone scales and cores may indicate a nearby cache. Caution should be exercised in collecting seeds or cones from caches, because of the danger of infestation by pathogenic fungi which may reduce germination.

Ants sometimes gather seeds together and in North Africa they have been observed to accumulate large piles of *Acacia* seeds. Any seeds collected from rodent or insect caches should be tested for soundness by cutting test or other means.

Seed Collection from Felled Trees

One method of collecting large amounts of seed is to synchronise it with normal commercial fellings carried out during the seed ripening season and to collect seeds or fruits from felled trees. If fruits are to be collected from throughout the felled crop, picking should for safety reasons be postponed until felling in the area is complete. If phenotypic quality of parent trees is more important than quantity of seed, it is preferable to select, mark and, if possible, fell and harvest fruits from superior mother trees in advance of the main felling. In *Pinus radiata* plantations due for clear-felling in New Zealand, a method has been used whereby a trained forest officer selected and marked the best 8 – 13 stems per hectare and the seed collection gang felled and trimmed the trees and collected the cones, so as to leave no obstructions for the subsequent felling of the remaining trees.

Collection of fruits from early thinnings should be avoided, since it is difficult to judge phenotypic quality correctly at that age. Felling crowns into existing gaps is advisable in order to facilitate recovery of cones. It is essential to confine collection to the season when seeds are mature; adjustment of felling dates to coincide with seed ripeness should be possible wherever the same authority is responsible for both the felling and the collecting activities, e.g. national forest services operating within state forests. Hand picking of cones or fruits in the fallen crowns is common practice, assisted by rakes, hooks or machetes. Small cluster-type cones, such as those of *Thuja* and *Tsuga,* may be harvested by cutting off the cone-laden branch tips and pulling them through a cone stripper. The stripper contains a series of teeth, similar to a rake, which are set sufficiently close together to remove the cones.

In practice collection in clear-felled areas has proved little if any cheaper than collecting from standing trees by a well-trained team of climbers, at least in north-temperate conifers. The tangle of fallen stems and crowns and the dispersal of some cones during felling greatly reduce productivity. Where operations are speedy and closely controlled, collection after the boles have been trimmed and removed but before the lop and top has been piled and burnt may prove the most efficient method.

Felling of selected individual trees specifically for seed collection may be necessary in areas where commercial fellings are not practised, for example when relatively small quantities of seed from a few trees are needed for provenance testing or other research purposes. Such special fellings should be avoided whenever possible, both because the bole is wasted and because the tree is lost as a future seed source, but they are sometimes inescapable in the case of tropical high forest species which are very difficult to climb and if a seed collecting expedition is severely limited in time.

Seed Collection from Standing Trees

By Hand

In the case of shrubs or low-branched trees, fruits can be picked directly from the branches by the collector while standing on the ground. Examples are *Crataegus, Sorbus* and *Ilex* spp. in temperate zones, the smaller acacias and mallee eucalypts in Australia and many of the small drought-resistant species of the arid and semi-arid zones. Smaller fruits are generally harvested directly into a basket, bag, bucket or other container held or worn by the picker.

Cutting, Breaking and Sawing

For branches out of arm's reach a variety of long-handled tools is available to enable the collector to reach the fruits from the ground. A pole and hook may be used to pull branches down within reach. Long-handled rakes, saws, chisels, hooks or pruning shears are used to pull off or sever individual fruits or fruit-bearing branchlets. Light rigid bamboo, aluminium or plastic poles 4 – 6 m in length are common. In order to reach beyond the 6 – 8 m range of single poles, multistage telescopic poles with a shear on the end have been developed. Robbins et al. have noted that, in some species, fruits or cones on the lowest branches may yield little seed, because of lack of pollination in that position, and that it is therefore preferable to collect fruits from at least half way up the crown. Ability to use long-handled tools efficiently from the ground is much affected by the density and form of the crown in individual trees.

A rope can be thrown or pulled over a branch as already described, but used to break off the seed-bearing branch rather than to shake it. A thicker rope is needed than for shaking. The method is not recommended for general use. It damages the tree, allows access to pests and diseases and, in the case of pines and other species which take two years to mature their seeds, destroys the next year's seed crop while collecting the current year's.

Several types of *flexible saw* have been used successfully to sever branches from the ground. One model, described by Anon. 1979, consists of a 3 ft. long flexible cutting cable fitted with precision-set carbon steel teeth and two 35 ft. polypropylene control lines. A sand-filled safety weight is used to project one of the control lines over the branch. An earlier model, the "commando saw", now no longer in production, was effective in servering eucalypt branches in Australia. Two operators could bring down branches up to 20 cm in diameter quickly and easily.

The method is not applicable to trees with acutely angled branches such as *E. tereticornis*. The other limitation depends on the efficiency of projecting the line over the desired branch. Rigid saws can also be used to sever branches. Sweney and Jones describe a method used in New Zealand in which a pruning saw or bow saw is attached by lugs to a 180 kg line running through a pulley previously raised into the crown. The pruning saw is used for small branches <2 cm diameter, while the bow saw will cut through a 10 cm branch in 5 minutes' sawing.

Use of Rifle

Another method of severing seed bearing branches is to shoot them down with a large calibre rifle. The method was successfully used to shoot out the tops of *Picea glauca* trees in seed production areas in north eastern USA. Not only was the topping of the trees found to be less expensive than climbing but the cones could also be collected at the best stage of development because of the short time in which the operation could be completed. More recently shooting off branches or tops from a helicopter has yielded promising results in Canada.

In Australia the collection of small samples of eucalypt and *Araucaria* seed from tall trees has been accomplished efficiently using a .222 or .243 or .308 calibre rifle with x4 telescopic sights. Branches up to 15 cm diameter could be brought down. "Pointed soft point" ammunition is more effective than "hollow point" when used with a .308 rifle.

A disadvantage of the rifle method is that very strict safety precautions must be observed. There are limitations to where a rifle may be used, for example not near roads or built-up areas. Also crowns of some species such as *Araucaria* and *Picea* may be considerably damaged by this technique.

In shooting down branches it is usually necessary to steady the rifle on a tripod or to rest the stock against a tree or the side of a vehicle. A clear line of sight is required and this can be a limiting factor in dense forests. It is usually best to shoot at right angles to the branch and to sever the bark on the underside with the first shot to avoid branch hang-ups. The bark on the upper side is then cut and finally shots are placed at intervals across the branch. It is important to select branches which will fall unobstructed to the ground. Horizontal branches are more readily detached than ascending branches. The shots should be positioned to take advantage of branch leverage. The method is best suited for collecting research quantities of seed from a heavy seed crop clustered on branches or tops too inaccessible to be conveniently reached by other means.

Seed Collection from Standing Trees

There is a limit to the height to which long-handled tools can be used for collecting seeds or fruits from the ground. Near that limit the operation consumes much time and energy but produces little seed. For tall trees which cannot be felled, therefore, climbing is often the only practical method of collecting. Some men are excellent natural climbers, while good training and good equipment can render collection by climbing an efficient and safe, albeit energetic, operation. For convenience the operation may be described under the following subheads: (a) Climbing into the crown by way of the bole, (b) Climbing into the crown directly, (c) Climbing and picking of fruits within the crown.

Climbing into the Crown by Way of the Bole

Climbing with minimum equipment: Climbing without mechanical aids is practised in a number of countries. In the Philippines some seed collectors climb barefooted or with the help of a rope which ties both feet together and presses them against the trunk of the tree. Other modifications are for the climber to cut successive notches in the bole with a hand-axe to support his feet, or to hammer in a series of iron spikes about 20 cm long which are later withdrawn for re-use as he descends. Both of these methods are physically exhausting, whether or not a safety belt is used, and do some damage to the tree. Climbing tall branchless boles with hands and feet involves a considerable safety hazard and the risks may tempt climbers to prefer collecting from the most easily climbable trees which are often silviculturally the least desirable. It is preferable to introduce one or other of the special climbing aids now available.

Climbing irons or spurs, which are attached to the climber's boots, offer a light and inexpensive means of safer and more efficient climbing, if combined with safety belt, strap and line, safety helmet of glass fiber and heavy leather gloves. The lightness of the spurs (less than 1 kg a set) makes them particularly suitable for use in inaccessible stands in roadless country, where all equipment must be carried on foot. They have been found to be the most efficient method for climbing trees of *Pinus kesiya* and *P. merkusii* in Thailand and for *P. caribaea* and *P. oocarpa* in Honduras and are in common use in many countries, especially for conifers.

There are a number of different types of climbing irons but basically they consist of a forged iron arm and connecting piece which terminates in a pointed spur. The iron must be fastened securely by a leather strap to the foot gear and sometimes to the leg of the climber. The spur may be of varying length but it is an advantage if the point does not extend beyond the sole of the boot, so that the climber can walk on the ground without difficulty. The optimum length of spur depends on the type of bark. 5 cm spurs are suitable for barkless telephone poles and thin-barked trees and are recommended for most species in Canada, while 9 cm spurs are better suited to species with soft, thick bark. Climbing irons should not be used when the bark is frozen and are not particularly safe on scaly bark.

Detailed guidance to the fitting, use and maintenance of treeclimbing spurs in Canada is contained in Yeatman and Nieman. The following description of climbing with spurs closely follows their account. The climber ascends the tree using a safety belt, with a safety strap or safety chain passed around the stem and hooked to the belt. A safety line is tied to the belt and two or more carabiners are clipped onto one of the belt rings. When climbing the stem, the climber must ensure that the spikes are well into the wood of the tree by keeping the knees out from the stem when setting the spur. The lower leg and ankle must be kept at a fair angle to the stem to prevent slipping and gouging the bark. The weight is kept on the feet spaced 15 – 20 cm apart and the centre of gravity away from the stem. The hands and arms are used to balance by holding the safety strap firmly in both hands, rythmically pulling the body towards the tree, moving the strap as the weight on it is lifted and tightening it in the new position as the body moves back. The pull on the safety strap is from the arms in ascending and it should not be transferred to the safety belt except when the climber is in a resting position. When the safety strap is tight, each foot is moved in turn and the weight transferred to the other foot. The safety strap is never unclipped except to bypass branches too heavy to break off. A second safety strap or carabiner with safety rope should be fastened above the obstructive branch before the first strap is unclipped. When sound branches are reached, preferably at the bottom of the live crown, the safety line is passed through a carabiner fastened by rope above the first branch, the safety strap is unclipped, and the climber works his way into the live branches.

The main disadvantage of spurs is the damage they do to the bark, particularly of thin-barked species. If climbing is only occasional, this should not be excessive, but frequent climbing of the same tree, e.g. for pollination and seed collection in seed orchards, is liable to cause an unacceptable degree of damage; other climbing methods should then be preferred.

Ladders: For heights from about 8 to 40 metres, vertical scaling ladders in several sections provide a safe and convenient means of climbing the bole to the live crown. They can be made of a variety of materials including wood, aluminium, magnesium alloy etc., but each section must be light enough to be easily pulled up by the climber. The legs of the bottom section can be placed on adjustable platforms for greater stability. The length of each section varies between 1.8 and 3 m and its weight should not exceed 3 – 4 kg.

The bottom one or two sections of the ladder are set up parallel to the tree stem with the bracket at the top against the trunk. The climber ascends with his safety strap around both the trunk and the ladder until his shoulders are level with the top of the ladder and then fastens it to the trunk by a rope or chain. Subsequent sections are pulled up by rope and fitted into the section below. Each section is climbed and fastened to the tree in turn. The same procedure is used for climbing from bole into crown as described above for climbing irons. With the lighter sectional ladders the climber can carry two sections, each 2 m long, attached to his safety belt. If the bottom two sections of 3 m each are fitted together and raised into position from the ground, this means he can erect a total of 10 m before needing to use the tool line to haul up additional sections.

Sectional ladders may be designed as "one-legged" or "two-legged". Two-legged are more common. The "one-legged" ladder comprises a central support with small bars fixed as rungs alternately on either side and is attached by chain or rope to the tree. The central support is usually made of steel or wood, so these ladders are not very different in weight from "two-legged" types. They have the advantage of being easier to lodge on uneven ground and they are also more easily manoeuvred between the branches and on very sinuous stems.

Sectional ladders can be used without any risk of damage to the tree. They can be awkward to handle in stands with dense canopy or undergrowth and are much heavier to carry than climbing irons, especially if long clear boles impose the need for many ladder sections. They are also more expensive. They are therefore of limited use in inaccessible roadless country, but are ideal in seed orchards or plantations in flat topography.

The Swiss tree bicycle or "Baumvelo" is a device for climbing tall straight trees which are branch-free to the live crown. It is lighter to transport than sectional ladders but heavier than climbing irons. It does no damage to the tree. It is suitable for use on stems with diameters ranging from 30 – 80 cm. Olesen found it very useful for climbing pine trees in Mexico. In the UK it was found particularly useful for conifer species with large cones e.g. *Pinus, Picea, Pseudotsuga,* in which the collector seeds to move from branch to branch to pull off the cones, rather than remaining for long in the same part. Its use in that country is now, however, restricted mainly to research collections, since the increased area of mature plantations allows bulk seed collections to be made easily from felled trees and ladders are more convenient for seed orchard collections.

The tree bicycle contains two separate units, one for each foot. Each unit consists of an arm (longer in the upper, shorter in the lower unit) to which is fixed a rubber supporting block which rests against the trunk. The lower end of the arm carries a stirrup or pedal with strap and quick release clips, which holds the climber's foot. The upper end is attached to a steel band forming a circle of adjustable diameter around the stem. The tree bicycle is used in conjunction with safety harness or belt, support chain or strap, safety line, safety clips or carabiners and nylon ropes.

Operation of the tree bicycle is well described by Seal et al. and Yeatman and Nieman. The steel band should not fit too tightly round the stem, which would hinder them sliding upwards. They will grip properly as long as the toe of the climber's boot does not touch the trunk when his full weight is on the stirrup; when that happens it is necessary to unclamp, shorten the loop and reclamp the band on each foot in turn.

After securing the bands round the base of the tree, the climber fits his feet into the stirrup, closes the quick-release clips and fastens the straps. He also passes his safety strap or chain round the stem, adjusting its length in accordance with stem taper. The climber ascends, putting his weight alternately on one or the other stirrup and lifting in turn the free foot to raise the loosened band. He pauses to adjust the bands, one at a time, to allow for the taper of the tree. Obstructing branches must be pruned flush with the stem as they are encountered.

The climber continues to ascend until the upper steel band meets the lower living branches of the crown. He then reaches into the crown to fit a nylon safety rope to hold the safety line and proceeds to "park" the tree bicycle. The essential operation is to tighten the lower band of the bicycle so that it grips the stem tightly even when there is no weight on it; there is then no risk of the bicycle slipping down the stem out of the climber's reach. The climber opens the ankle clips on both stirrups, frees his feet from the straps, unhooks his safety strap and climbs into the crown.

The tree bicycle provides an extremely safe means of climbing straight branchless trees without damaging them and is lighter and more portable than sectional ladders. Some practice is needed but most men become competent and quick in its use within a few days. Its main disadvantages are the cost, the fact that its use is limited to a certain range of diameters and that, unlike ladders and climbing irons, it requires the bole to be pruned of branches all round the circumference up to the living crown. Where regular climbing of the same trees is foreseen, however, as in seed stands or seed orchards, the cost of thorough initial pruning is fully justified. It is advisable to work always with two tree-bicycles within call of each other, because a man in difficulties on a tree at a height greater than can be reached by a ladder can *only* be reached on another tree-bicycle, or by climbing irons.

Direct Climbing into the Crown

Ladders: Access to stout lower branches in the crown may be obtained directly from the ground or by ladder, provided that the branches are not too high. Free-standing domestic step-ladders or taller tripod ladders have the advantage that they do not need to be rested against the tree; they are awkward to handle in dense stands but are suitable for collection in seed orchards or heavily thinned plantations where trees are widely spaced. Tripod ladders up to about 6 m are available but must be guyed with ropes on each side for stability.

General purpose ladders are available in wood or aluminium alloy and may either be constructed as a single unit or may be extending, when they consist of two or more sections constructed so that the height may be varied by a relative sliding movement of the sections. Single-section ladders designed for fruit-picking have a splayed base fitted with moulded rubber feet or metal spikes to give a firm grip on soft ground and can be used to heights of 8 – 11 m.

The Finnish Tarra-tikkaat ladder, designed specially for forestry, is similar but incorporates an aluminium ring which slides up and down the ladder, can be attached to the climber's belt and locks automatically if the climber should fall. It also includes a strut attached to the ladder which grips the stem and maintains a distance of about a meter between the bole and the top of the ladder. This enables the picker to reach further out into the crowns of small trees of species such as *Cupressus* which bear fruits near the ends of the branches.

General purpose single-section ladders extend up to about 11 m, while doublesection extending ladders extend to about 14 m. They may be leaned against a stout branch or against the bole for direct access to the crown. The top must be made fast to the tree with a nylon rope and a tall ladder must be further steadied with two guy-ropes. Depending on species and crown form, the climber may leave the ladder and climb into the crown, using safety belt, safety line and nylon safety ropes as from the tree bicycle, or he may collect fruits while standing on the ladder; in the latter case he should secure himself to the ladder by means of a nylon safety rope, while foot fatique can be reduced by a movable foot stand that can be hooked on to a convenient rung.

The larger and heavier ladders need a vehicle for transport between stands. Extension ladders mounted on trucks can be used to reach crowns of up to 20 – 25 m, but the lack of manoeuverability restricts the vehicle to roads or very flat ground. A light alloy ladder mounted on a tractor is described by Morandini; it will reach to 14 m and the tractor can be used in forests without roads. In the USA a magnesium extension ladder mounted on a halfton 4-wheel truck, can be used with safety to 9 m. Sway from the vehicle suspension is eliminated by the use of four mechanical jacks attached to the support posts of the ladder platform rack.

Climbing and Picking Fruits within the Crown

Methods of climbing and picking fruits in the crown are independent of the method used to reach the crown, whether this is by ladder, tree bicycle or climbing irons. The following summary of the main points to be observed is based on Yeatman and Nieman.

In climbing on the branches of a tree, as on the rungs of a ladder, handholds are used mainly for guidance and balance and the feet and legs are used for thrust. Only one limb - a hand or a foot - is moved at a time, gripping or standing on branches where they are strongest, close to the main stem. Doubtful branches should be tested by being pulled sharply before being trusted to carry the weight of the climber. Confidence and muscular coordination are keys to safe climbing. A sustained state of tension and reliance on strength alone can be both exhausting and hazardous. Most of the work should be borne by the legs.

Tension in a climber may be greatly reduced if he has confidence in a proven safety system. Occasional mishaps may occur even to the best climber, if only because a branch which appears safe turns out to be treacherously brittle or slippery. Modern climbing techniques and equipment are designed to ensure that, if a climber falls, he falls only a short distance before his fall is arrested.

Excellent illustrated accounts are available in Seal et al., Yeatman and Nieman and Robbins et al. and include details of the ropes and knots which the climber must know how to use.

Safety belt and strap: One essential item of equipment is the safety belt or harness. The most convenient type consists of a waist belt, which carries a number of metal rings to which may be affixed the safety line, safety strap or climbing chain, short lengths of safety rope, tool line etc., and a saddle which fits beneath the buttocks, is connected to the waist belt and can be used as a seat for the climber when this is convenient for collecting. Additional support can be provided by including a pair of shoulder straps as part of the harness. Attached to the safety belt are the equally essential safety strap or chain and one or more short nylon safety ropes which serve to secure the climber to the stem of the tree. One end is fastened to one side of the belt while the other is passed round the stem and then attached to the other side of the belt.

A light chain is sometimes used as the climber ascends the bole, the length of which can be readily adjusted to the taper of the tree by taking in a few links. An adjustable safety strap is as good or better. A second, reserve short rope should be carried to enable a climber using climbing irons or ladders to surmount an obstructing branch below the live crown; the reserve rope is fitted above the branch before the original one is unfastened below it. The safety rope is also fitted to the upper stem while the climber picks fruits within easy reach. Thus safety belt and safety ropes together will, on suitable trees, ensure the safety of a climber as he ascends the bole and while he picks fruits, even in the absence of a safety line. But they do not ensure safety as he climbs within the crown.

Safety line and safety ropes: Another important item is the safety line. Nylon rope of 12 – 14 mm diameter and weighing abour 1 kg per 10 m is commonly used. Apart from its strength and good wearing properties, it has advantage of being somewhat elastic, thus reducing the jarring effect on a climber's body as the rope tautens after a fall. Polypropylene is less suitable since it melts at relatively low temperatures. However, it has been used successfully in some countries e.g. Honduras, where nylon is inobtainable. Precautions taken are to use polypropylene rope at least 18 mm in diameter and to avoid any techniques which would involve prolonged friction on a stationary rope. Temporary friction of a moving rope, e.g. over a branch, is not dangerous since the heat generated at any one point in the rope is very little. Natural fibres such as hemp or jute need to be substantially heavier than nylon if they are to attain the same strength, in addition they are subject to rot especially if used in wet weather.

Safety lines or ropes may be conveniently divided into two classes, the short safety rope which the climber uses to attach himself to the tree while working in the crown and the long safety line which runs down to the ground and is controlled by an anchorman. The short safety rope varies in length from 3 to 10 m and is of the same gauge as the long safety line; a climber will normally carry two of them attached to his safety belt. When he is working in the crown, he attaches one of them either to a stout branch or to the stem. If he is in the upper crown where the stem is too thin to be safe, he attaches the rope to the stouter portion of the stem *below* him and passes it spirally upwards round the stem. If the stem is stout enough and he uses a safety rope of appropriate length, he can reach well out into the outer crown. The short safety rope is simple to use and renders the climber completely independent while collecting in the crown.

One end of the long safety line is attached to the climber's safety harness, the other is controlled by an anchorman who remains on the ground. The length of the line must be at least twice the height of the tree, to allow a climber to be lowered to the ground, if necessary, from any position close to the top of the tree where he may be working. Apart from the additional safety factor it provides, the safety line can enable the climber to reach fruits in the outer crown which would otherwise be inaccessible.

Method of climbing with use of safety line: On reaching the lower crown, the climber ascends in a slight spiral round the stem, so that he takes his long safety line across branches which will hold the line in the event of a fall. If live branches are lacking on one side of the stem, the climber should follow a zig-zag course on the side of the stem bearing live branches. If stout branches are lacking, the climber should fit a safety rope to the stem and clip the safety line to it. The aim should be to ensure that he cannot fall freely more than 2 m before the safety line starts to take his weight.

Meanwhile the anchorman, as described by Yeatman and Nieman, must stand well clear of the fall line from the tree being climbed to avoid possible injury from falling branches that may be broken off as the climber ascends. The anchorman must be ready at any time to support the weight of the climber on the safety line as he feeds it under one arm, around his back, and over the other shoulder, releasing it alternately with one hand while gripping it with the other. A half turn of the rope should be made around the lower trunk of a neighbouring tree to provide additional friction and greater security should the climber slip or fall.

However, the friction should be minimal during the ascent to ensure that the climber is not held back by the safety line. It is important that the unused part of the safety line remains coiled on the ground and free of obstruction so that tangles are avoided as the rope is paid out. The long safety line is of particular advantage when it is necessary to take rapid evasive action from wasps or bees that are sometimes encountered in tree climbing.

When the climber reaches the level in the crown where he is to collect, he may secure himself by passing the safety line over a stout branch and around the stem. Alternatively he may secure himself to the stem by means of a safety rope passing it over a stout branch and around the stem above him and re-attaching it to his safety belt then detaching the safety line. If the branches are not stout enough, he fixes a short safety rope to the stem as high above him as possible and passes the safety line through it. The rope should never be attached to the stem above the point where the diameter falls below 8 – 10 cm.

The climber can now either go on to pick the fruits with the anchorman still holding the safety-line or he can apply a special safety line locking method, using carabiners or safety clips, and take charge of his own movements. In either case most of his weight can be supported by the harness saddle, enabling him to reach further out in the crown than would be the case if his entire weight had to be supported on relatively thin branches. Use of the safety line locking method allows the climber to make a controlled descent as he works downwards collecting fruits as he goes.

Method of descent: For descent the climber must first climb up and release the topmost safety rope if one was used, while the anchorman takes in the slack of the safety line. Then, as described by Yeatman and Nieman, the climber descends with care, following the same route down as the one he ascended by, so as not to foul his safety line on the small branches of the upper crown. When his feet reach branches that will support all his weight on the safety line and are large enough in diameter to withstand rope abrasion, he moves laterally to catch the safety line on the crotch of a branch.

The climber continues down so that the trailing safety line passes up and around the stem, over a stout branch, and on around and down to the climber. Alternatively, in trees with dense crowns, it is often easier to stop, attach oneself to the tree, untie the safety line, pass it up on one side of the trunk over a branch on the opposite side and down the other side, and retie it to the body belt. With the safety line securely caught on a branch (crotched), the climber may be let down by the anchorman paying out the rope or he may control his own descent by using the safety line locking method.

Descent to the ground by safety line is the best method when the ascent of the bole has been by climbing irons. When ascent has been by tree bicycle or sectional ladder it is usually preferable to descend the some way; otherwise the climber must loosen the bands of the tree bicycle or disconnect the sections of ladder while suspended from the safety line.

Method of picking fruits: With correct use of the safety harness and safety line the climber should have both hands free to pick the fruits. Methods vary according to the size, number and distribution of fruits and the firmness of the peduncular attachment. Numerous, small, clustered and accessible fruits can be picked and deposited immediately in a bag attached to the climber's belt or slung from his shoulder and with its mouth braced open. Bigger and more scattered fruits may be detached, dropped and collected from the ground later. Fruits too inaccessible to be picked by hand may be detached by long-handled pole, hook, rake or shears. Tools used in the crown should be pulled up by the climber by the use of a light tool line attached to his belt, after he has reached the collecting area in the crown.

In Chile climbers beat the branches with a long stick to detach fruits of *Nothofagus alpina*, but the method is less effective with *N. pumilio*. Some form of hook is in common use for conifers. In Thailand the hooks used for collecting cones of *Pinus kesiya* and *P. merkusii* have been designed with two sharpened edges pointing in opposite directions, so that they can be used to detach cones with either a pushing or a pulling action, thus avoiding the breakage of twigs with new conelets. They are mounted on light steel tubes which can be extended to a length of about 3 m. Various types of hook, rake and shears are illustrated by Robbins et al. Gathering of fruits from the ground, when they have fallen after detachment by tools, is the same as described above for natural seed fall.

If fruits are small, numerous, clustered, inaccessible for hand picking or held firmly by the peduncle, it may be necessary to cut off whole fruit-bearing branchlets by long-handled shears or saw. *Pinus oocarpa* is an example of a species in which the cones have strong woody peduncles which are difficult to sever without damaging the branchlet. In such cases one alternative is to collect by cutting off the branchlets, but this causes destruction of the next year's crop of conelets. A better solution is to devise new equipment to overcome special local problems. In Honduras a bell-shaped sharpened hook is designed to be pushed outwards from the centre of the crown towards the ends of the branches and this severs the cones while doing the minimum damage through bending or breaking the branches. The hook is mounted on a 5 m aluminium or wooden handle. It has been in use for over 5 years and has proved to be very practical for cone harvesting from *P. oocarpa, P. caribaea, P. patula* subsp. *tecunumanii* and *P. maximinoi.*

If fruits are clustered at the end of long branches where they are out of reach of the climber, he may need to cut off the whole branch. In Brazil climbers cut off fruit bearing branches of *Eucalyptus* with a knife on a pole and the usual practice is to collect from alternate trees in successive years so that the tree crowns have a chance to recover.

Climbers should wear boots with non-slip soles and overalls without belts or loops which could catch on branch snags. A close-fitting safety helmet, gloves and safety glasses are often useful. Anchormen must wear industrial safety helmets to protect them from objects falling from the trees. A light pruning saw is often necessary to provide a clean bole suitable for climbing by tree bicycle.

The methods described above have proved highly reliable when used on coniferous species. There is less experience and probably greater difficulty in climbing hardwood species of the tropical moist forests, where a few massive branches separated by intervals of 3 – 4 m pose a special problem. Very thorny species such as *Acacia* spp. present another problem but, because they rarely exceed 15 – 20 m in height, collection can often be made from the ground.

Useful aids to collection described by Doran et al. include a strong vehicle roof carrier to stand on, light ladders fitting on the roof carrier, a rope and weight for throwing over a branch to shake or break it, flexible saw or long-handled pruning shears and thick leather gloves for stripping pods from thorny species. Standard collection methods will certainly need to be modified to adapt them to the peculiarities of individual tropical species.

SELECTION OF SEEDS

Good quality seeds are essential to grow a strong and healthy crop. Healthy seeds can be bought from trusted sources or farmers can produce their own seeds. In that case, seed selection can be used to improve the quality of seeds. There are several diseases that are transmitted via the seeds. If seeds from an infested field are used to grow the next crop, these seed-born diseases will immediately cause serious problems. Seed selection should thus start by obtaining seeds from healthy plants. Small, shriveled and broken seeds contain less nutrition for the developing seedling. By removing these inferior seeds, the farmer is able to grow stronger and healthier seedlings. Even if seeds are selected carefully, they still may carry diseases. In these cases seed treatment (e.g. hot water treatment or chemical seed treatment) could be used to further improve the quality of the seeds.

While seed selection is mainly aimed at obtaining healthier seeds, it can be used also to maintain and improve the quality of the crop variety. In a crop field, there are always differences between plants. Some plants may have characteristics that are more suitable than those of other plants. During the growing season, the farmer can try to observe these differences and mark preferred plants with a ribbon or with a stick. During the harvest, the seeds of these plants can be reserved for growing the next crop. In this way, the farmer can slowly improve the quality of his variety. The selection of these plants may be based on characteristics such as the size of the plant, color or size of fruits, number of grains per ear, etc. But selection can also be done to keep seeds of plants that suffered less attacks by insects or diseases. In this way, the farmer will select at harvest time the best seeds to be kept apart for the next season. Still, just before starting the next season, it is recommended to select once more to remove seeds that are too small, spotted, deformed, discolored, etc. For sowing only the very best seeds should be used.

Forage Seed Quality

Forage seed quality depends on three key components:

- germination,
- genetic purity, and
- crop purity.

Germination

Germination is the measure of how well a seed produces a normal, healthy seedling when planted under ideal conditions. The standard germination test, evaluated under the temperature considered ideal for each species, provides an estimation of seedling emergence if soil conditions at planting and during early seedling growth are near optimum. The results of the standard germination test are printed on the seed tag.Each species has a specific range of temperatures at which germination occurs. If temperature at planting falls outside this range, germination may be slowed or prevented, and during extreme conditions the seedlings may die. Excessively high or low temperatures also can induce secondary seed dormancy, causing a delay in germination and establishment. It is therefore important for growers to be aware of the germination requirements of the species that is planted and to sow fields when conditions are favorable. It is essential that forage seed have adequate moisture immediately following planting to promote rapid germination and for a period of about three weeks after emergence to promote seedling growth.

Genetic Purity

Varieties are adapted for specific uses and production areas. The genetic constitution of a variety will influence disease and insect resistance, plant response to climatic extremes, forage quality, and many agronomic characteristics. Growers should purchase seed of a known variety that has been tested for their area. Genetic purity, or trueness to variety, is established and maintained by special purification and seed increase programs, by field and seed inspections, and by pedigree records. The best assurance of obtaining genetically pure seed is to buy certified seed, clearly identified by the "blue tag". The blue certified seed tag is issued only to seed lots that meet specific genetic, physical, and physiological standards.

Crop Purity

Seed lots with even small amounts of weed seed or other crop seed can cause serious economic losses and diminish forage quality. Weeds and other crop plants compete with the desired species for nutrients, space, and soil moisture and are often difficult or expensive to control. Some weeds, such as serrated tussock, Canada thistle, or horsenettle, can harm animals or threaten pasture value. Other weeds that are frequently found in forage seed include dock, sedge, wild garlic, wild onion, mustards, panicums, and other forage seeds such as ryegrass.

STORING SEED

High-quality forage seed can be stored successfully for more than one season if it is kept in a cool, dry environment. In general, the shelf life of grass seed stored at room temperature (77°F) is roughly two years. Orchardgrass and tall fescue, however, are exceptions and can be stored for only about one year. Legume seed storage potential is fairly good; most can maintain germination for two years, with the exception of crimson clover, which can be stored only one year.

Avoid storing seeds in metal storage buildings or attics. Temperatures there can be extremely high during the summer, and it does not take long under these conditions for germination potential to be destroyed. Basements are well suited for seed storage as long as moisture can be controlled.

Forage seeds may be stored in freezers; however, freezing temperatures may kill Endophyte and Rhizobia. Seeds are a tempting food source for rodents and many insects. To protect seeds from rodents, store them in metal trash cans with the lid securely fixed. To control insects, tape no-pest strips on the inside of the lids. Do not insert the strip into the seed bag. Replace the strips every two to three months.

After storage, it is essential to test seed for germination at least one month before anticipated use. This will allow time to locate fresh seed if germination has declined to low levels during storage. The seed lab also can test seed lot viability using the tetrazolium chloride (TZ) test. TZ tests can be performed in 24 hours, giving growers an accurate and rapid evaluation of seed lot quality. There is a small charge for TZ tests.

SEEDING RATE

Seeding rates should be based on the amount of pure live seed (PLS) in the seed lot. This is particularly true for chaffy native species with fluffy seed coats or large awns or for seed that are difficult to clean. In the latter case, the seed contain large amounts of inert matter, which makes calculating seeding rates difficult. Suggesting seeding rates based on bulk seed weights can result in low plant populations. Common species like fescue, ryegrass, orchardgrass, and legumes generally have 95 percent or greater purity with germinations above 90 percent. Therefore, it is customary to suggest planting rates based on bulk seed (pounds of seed per acre).

PLS is calculated by multiplying seed lot germination by purity. For example, if germination is 90 percent and pure seed 90 percent, PLS is 81 percent [(90 x 90) ¸ 100]. The actual seeding rate based on PLS can be calculated by dividing the recommended rate by the PLS. If the recommended seeding rate is 15 pounds per acre, the actual seeding rate would be: 15 pounds per acre ¸ 0.81 = 18.5 pounds per acre.

Seed-testing facility perform an accurate evaluation of germination because a trained seed analyst can best judge germination potential. Many forage species also require special germination-test environments not readily available to growers. Some freshly harvested species require special pre-chill treatments before they will germinate. It is best to have a reliable evaluation of fresh seed performed by a qualified laboratory.

REFERENCES

Baskin, C.C. and Baskin, J.M. (2018) *Seeds: Ecology, biogeography, and evolution of dormancy and germination.*San Diego, Academic Press.

Bewley, J. Derek, and Michael Black. 2014. *Seeds physiology of development and germination. The language of science.* New York: Plenum Press. page 230.

Gutterman, Y. (2022) *Seed germination in desert plants.* Springer Verlag, Berlin/ Heidelberg.

International Workshop on Seeds, and G. Nicolas. 2023. *The biology of seeds recent research advances : proceedings of the Seventh International Workshop on Seeds, Salamanca, Spain 2002.* Wallingford, Oxon, UK: CABI Pub. Page 113.

2

Wheat Seed Production

Wheat is a high-volume, low-profit seed crop and has been produced primarily by heavily subsidised government seed programmes. With privatisation and liberalisation, many of these programmes are at risk of being closed down. The private sector, however, may not focus on wheat seed due to its characteristics (self-pollinating, high-volume and low-profit). If private seed enterprises exist, they consider wheat seed to be of secondary importance. Furthermore, in most countries there has been no ongoing effort to promote the use of improved seed by wheat farmers, and no significant breeding developments have recently taken place to increase yield and quality. Since wheat is a self-pollinating crop and the grain can be used as seed, farmers tend to replant their own seed. It is, therefore, expected that in the future the large majority of resource-poor, small-scale farmers in many developing countries will have to rely on seed saved from the previous harvest.

In the developed world, declining world market prices for grain do not encourage farmers to produce wheat. Moreover, there have been no widespread disease outbreaks, which make seed treatment (and thus purchase of certified seed) necessary. Countries in Europe and the United States are attempting to reduce subsidies, which will result in further unwillingness to invest.

On the other hand, there are possibilities in developing countries for major increases if significantly improved varieties appear on the market, disease outbreaks occur and organised promotional efforts emphasise maximum production efficiency including improved seed. Introduction of Plant Variety Protection and/or sophisticated technologies, such as Genetic Use Restriction Technology (GURT), may also induce the private sector to invest in wheat seed.

In Australia, Europe and the United States, there is a possibility of increased use of seed of higher yielding varieties if subsidy reductions and/or market trends cause a decline in production area, with only the more efficient farmers remaining in production and attempting to produce more (or the same total amount) and more efficiently from less area as a result of trying to diversify crop production.

WHEAT SEED

Seed is a means of dispersal for plant populations in space (spatial) and time (temporal), representing continuity and change, and thus adaptation to the local environment. Seed has played a critical role in agricultural development since prehistoric humans domesticated the first crops. In modern agriculture, seed is a vehicle to deliver almost all agriculture-based technological innovations to farmers so that they can exploit the genetic potential of new varieties. The availability, access and use of seed of adaptable modern varieties is, therefore, determinant to the efficiency and productivity of other packages (irrigation, fertilisers, pesticides) in increasing crop production to enhance food security and alleviating rural poverty in developing countries.

Figure 1. Wheat seeds

For seed to play a catalytic role, it should reach farmers in a good quality state, i.e. high genetic purity and identity, as well as high physical, physiological and health quality. In contrast to fertilisers and pesticides, farmers select and save seed to plant the next year's crop, and any off-farm seed from the formal sector should be of a better quality for farmers to invest in it. Therefore, the best production techniques need to be followed to produce good quality seed.

For wheat, seed and grain production follow rather similar operations but different strategies. Apart from good agronomic management of the crop, seed production differs from grain production on the following key issues: land requirement, isolation, roguing, prevention of contamination and limitations of generations. Another difference is that seed crops must meet specific quality standards prescribed by the national seed regulations. The technical, administrative and legislative control by the certification agency provides guidelines that have to be followed to produce good quality seed that meets the standards.

VARIETY DEVELOPMENT, EVALUATION AND RELEASE

Modern varieties are the backbone of the formal seed industry. National Agricultural Research Systems (NARSs) have a major responsibility for variety development and for generating appropriate technologies to better utilise the yield potential of new varieties. The new varieties must pass through a series of evaluation, release and registration tests and procedures before farmers can use them. Basically two type of tests are carried out: Distinctness, Uniformity and Stability (DUS) and Value for Cultivation and Use (VCU).

A DUS test is a descriptive assessment that establishes the identity of the new variety, by using morphological characters, as well as its uniformity and stability. It is a useful tool for the purposes of seed production, certification and plant variety protection. The DUS tests usually run for two years. The new variety is compared with existing varieties to establish its distinctness. A variety description is prepared, and differences with other varieties noted.

The VCU trials focus on the benefit of the variety to the end users, farmers and consumers. National multilocation variety trials enable identification of superior varieties that meet diverse agronomic and consumer requirements. The VCU tests usually run for three years. In some countries, the variety is tested in on-farm verification trials under farmers' management conditions during the last year.

Based on DUS and VCU test results, a variety may be released and registered for farmers' use. Many developing countries give priority to agronomic (VCU) trials rather than descriptive (DUS) tests. While both tests are important, the benefits of the two tests must be considered based on the immediate need of the country to use available resources efficiently and economically.

SEED CLASSES

Seed production follows a generation system to ensure that all seed that is marketed to farmers originates from a known source (breeder seed). When a variety is officially released, the small amount of breeder seed received from the breeder (agricultural research centre) is multiplied through a number of generations before it becomes available to the farmers in larger quantities as certified seed. Each generation is produced under strict supervision and must meet seed quality standards. The number of generations that are allowed after breeder seed depends on the mode of reproduction of the crop, risk of contamination, multiplication ratio and quantity of the seed required. For wheat, four to five generations are commonly used.

Different generation schemes exist (Table1), which vary very little, particularly in nomenclature. The procedures followed are essentially the same.

Table 1. Comparative seed nomenclature in selected countries of West Asia and North Africa

Definition	OECD	AOSCA	Ethiopia	Egypt	Jordan	Algeria
1st generation supplied by plant breeders	Breeder	Breeder	Breeder	Breeder	G_1	G_0 (G_1, G_2)
2nd generation	Pre-basic	Foundation	Pre-basic	Foundation	G_2	G_3
3rd generation	Basic	Registered	Basic	Registered	G_3	G_4
4th generation	Certified	1 Certified	Certified 1	Certified	Certified 1	Registered 1
5th generation	Certified 2	—	Certified 2	—	Certified 2	Registered 2

Source: FAO, 1975; van Gastel and Hopkins, 1988.

1. Breeder seed is the initial source of seed and is usually produced by the breeder. It the source for the production of pre-basic or basic seed.
2. Pre-basic seed is the progeny of the breeder seed and is usually produced under the supervision of a breeder or his designated agency. This generation is commonly used for crops that have low multiplication ratios and where large quantities of certified seed are required.
3. Basic seed is the progeny of breeder or prebasic seed and is usually produced under the supervision of a breeder or his designated agency and under the control of a seed quality control agency.
4. Certified seed is the progeny of basic seed and is produced on contract with selected seed growers under the supervision of the seed enterprise, public or private. Certified seed can be used to produce further generations of certified seed or can be planted by farmers for grain production.

Breeder seed production is not monitored by the seed certification agency, while basic seed and certified seed are covered in the seed certification scheme. The seed quality control agency verifies the quality both in the field and in the laboratory and certifies that the seed meets the national standards. Such classes of seed are known as certified. It is important to note that all certified seed classes relate to a breeder seed through one or more generations.

Some developing countries, where natural disasters such as drought are a common phenomenon, recognise a 'commercial seed' class to meet seed shortages in emergency situations. In such cases, the standards for certified seed are often lowered and accepted for distribution to farmers to overcome seed shortages. In other situations, commercial seed is simply a grain used as seed after laboratory testing for some quality attributes, such as purity and germination.

Variety Maintenance and Breeder Seed Production

Upon release of a new variety, a breeder will make available a small quantity of seed stock that is very pure and represents the variety. This stock is referred to as parental material and forms the basis of any future maintenance and seed multiplication of the variety. Laverack defines maintenance as "the perpetuation of a small stock of parental material through repeated multiplication following a precise procedure". For wheat, an ear-to-row (Plate 74) method is recommended, where a number of ears (depending on the total quantity of certified seed required) that are true-to-type are selected, threshed separately and then planted in individual rows. During the entire growth period, the rows are inspected regularly, and any row with off-types or deviants is discarded. Ears are selected from the remaining rows to repeat the cycle, which is usually referred to as maintenance. The remaining rows are bulk harvested, and the seed is called breeder seed. In India, for example, 1 000 plants are used for wheat variety maintenance.

Each year the cycle is repeated to provide a regular supply of breeder seed for further multiplication to basic seed and then to certified seed. Maintenance and breeder seed production is the responsibility of the breeder or the institution that developed the variety. In many developing countries, maintenance is seldom carried out properly, and the responsibility is often taken over by seed production organisations. Some national seed programmes (e.g. Ethiopia) have established special farms to produce early generations (pre-basic and basic seed) to maintain quality and availability.

Laverack described different arrangements and management approaches for breeder seed production that can be useful and adopted in developing countries. In Morocco, a separate Seed Unit has been established within the Institute National de la Recherche Agronomique to maintain and produce breeder seed of public varieties.

PRODUCTION PLANNING

New improved varieties developed by NARSs should be multiplied and made available to farmers in the shortest possible time to realise the benefits of investments in agricultural research. Appropriate seed production techniques coupled with strict quality control measures ensure that varietal purity and identity is maintained, which is the cornerstone of the entire seed programme. The rate at which the variety is multiplied and accessed restricts the availability of seed and its adoption and rapid diffusion through formal or informal channels.

Most farmers have a tendency to save their own seed of self-pollinating crops such as wheat compared to open-pollinated crops or hybrids where the risks of contamination and decrease in yield are higher. The seed renewal rate, the frequency in which farmers purchase certified seed of the same variety from the formal sector, varies between the type of seeds, crops, production environments and socioeconomic factors.

The multiplication factor can be defined as the amount of seed harvested from each quantity of seed sown. It differs from crop to crop, but is largely dependent on climate, physical factors (soil fertility, etc.) and agronomic management.

New varieties, after they enter commercial production, may lose their genetic potential or become susceptible to pests over time, which requires their replacement. Moreover, the varieties may also be exposed to genetic, mechanical and pathological contamination during the seed multiplication process. There is a practical need to limit the number of generations that the seed is multiplied after breeder seed.

For each generation, it is important to make an accurate estimate of the seed that needs to be produced every year. The total requirement depends on: (i) crop area planted; (ii) seed renewal rate; (iii) seed rate; (iv) multiplication ratio; and (v) number of generations. When planning seed production, the rates of rejection during field inspection, processing and laboratory testing should also be considered. Furthermore, a certain amount of carry-over seed should be taken into account, particularly for early generation seed. Precise estimates of seed demand are difficult because several factors, such as weather conditions and socio-economic factors, influence farmers' decisions on which crop to grow the next year. In case of wheat, farmers may prefer to keep own-saved seed for a longer period than the anticipated seed renewal rates of four to five years. The seed demand of new varieties is even more difficult to estimate and depends on the adoption rate, which in turn is influenced by several technical, institutional, economical and sociological factors.

PRODUCTION ARRANGEMENTS

Generally, early generations (breeder and pre-basic seed) are produced on agricultural research farms and basic seed on specialised seed farms, while certified seed is more conveniently and economically produced by farmers who are contracted by the seed enterprise (public or private). Farmers' cooperatives and associations are often involved in certified seed production. The seed enterprise provides technical guidance and advice to growers on various aspects of seed production, such as planting, roguing and harvesting, as well as monitoring and supervising field operations. The seed quality control agency inspects the fields and tests the seed in the laboratory.

In wheat, seed and grain crop production follow rather similar operations. The end product can either be used for consumption or for planting. Such similarity can easily lead to mixing seed and grain during planting, harvesting, transportation and storage. Seed production should be strictly monitored, and seed producers should be quality conscious. In developing countries, farmers are rigorously selected before they become seed growers.

In many developing countries, the availability of experienced seed growers is limited, and seed organisations provide considerable support. A village seed production scheme, which groups adjacent fields for seed production, will help to overcome problems of isolation distances and supervision when contracting smallholder farmers for seed production.

The cost of seed production is higher than the cost of producing grain for consumption because it involves extra operations to maintain quality. Thus seed growers incur additional costs in field operations and management and must be paid a premium. In case of wheat, the additional premium paid over the grain price ranges from 10 to 20 percent.

Land Selection

Wheat can be successfully grown in most parts of the world, both in tropical and temperate environments and on all soil types that are well drained and productive. Seed (certainly breeder seed) should be produced in areas: (i) where the variety is adapted; (ii) where soil conditions are optimal (to achieve a high multiplication ratio); and (iii) where climatic conditions are reliable to avoid loss due to natural hazards (flooding, drought, frost, etc.). The early generations should possibly be planted at two different locations to reduce the risk of losing the complete generation.

In some tropical countries, there is a high risk of rain coinciding with harvest time. Rainfall at maturity delays harvesting, causes sprouting and predisposes the crop to fungal attack, resulting in poor seed quality, i.e. reduced viability and vigour. Selection of appropriate sites is very essential to produce quality seed.

Previous cropping

The crop should be planted on a field with a known history to avoid contamination from volunteer plants, noxious weeds and soilborne diseases that are potentially seed transmitted. The minimum number of years that is allowed between the planting of two seed crops (or seed and grain crop) is usually prescribed by the national seed regulations. Examples of requirements for previous cropping in countries in West Asia and North Africa are provided in the WANA Secretariat.

A wheat seed crop should never immediately follow wheat, unless the wheat crop in the previous season was of the same variety and of the same or higher generation. For basic seed, it is often recommended that the field should not have been planted with wheat or other small grain crops for at least two consecutive years. For certified seed, no wheat crop should have been grown in the previous year.

A suitable crop rotation plays an important role in pure seed production. For example, a two-year rotation for flag smut and seed gall nematode is suggested where applicable. For wheat, previous cropping could be legumes, vegetables or clean fallow, but other cereals (barley, oat, rye and triticale) and forage crops (oat) should be avoided.

Weeds

Seed contaminated with weeds could be the means for introduction and dissemination of noxious weeds. Mohamed (1996) mentioned that the introduction of wild oats in wheat in Egypt is attributed to contaminated seeds. Therefore, heavily infested fields with noxious weeds should be avoided. Unfortunately, fields in many countries are infested with wild oats (*Avena fatua, A. sterilis* and *A. ludoviciana*), noxious weeds that are spread almost all over the world and are difficult to eradicate. *Avena fatua* has no resistance to frost, and it is only problematic in summer wheat. In West Asia and North Africa (WANA), the problem is mainly *A. sterilis*. Worldwide, a 10 percent potential yield loss in cereals is due to weeds, even with currently used control measures. In wheat, Tanner and Sahle cited that the use of clean seed had a significant effect on grain yield compared to weeding by hand or hoe.For land selection, the most essential prerequisite is the selection of clean fields that are properly rotated and known to be free from contaminants.

Planting

Seedbed

Seedbed preparation is the same as for a grain crop. Wheat does not suppress weeds sufficiently and needs a clean, weed-free seedbed for planting. In Sudan, weed competition studies on wheat showed that if weeding is not carried out between two and four weeks after sowing the yield is reduced by 20 percent.

Planting method

Planting with an automatic drill is recommended but not essential. However, row-planting has an advantage over broadcasting, as it requires less seed and facilitates mechanised weed control, roguing and field inspection. Roguing lanes (empty rows at intervals) should be left, which could be used by the seed grower to walk through the field when roguing and inspecting the crop, as well as for spraying the crop. Deep sowing delays emergence, resulting in weaker seedlings, reduced emergence and poor tillering and yield. Varieties with short coleoptile length, particularly semidwarf varieties, suffer most compared to varieties with longer coleoptile length.

Seed rates

The optimum seed rates for wheat vary with variety, location and method of planting. For seed production fields, a lower seed rate may be recommended because lower seed rates lead to higher multiplication factors but to lower yield per unit area (Table 2). Higher multiplication factors lead to rapid seed increase (more seed harvested per kilogram of seed planted), and farmers will benefit from the improved variety earlier. Low seed rates do not only increase the multiplication factor, but also often improve seed quality because a lower number of plants per unit of land receive better nutrition, thus producing better quality seed.

Table 2. Effect of seed rate on multiplication factor of wheat

Seed rate (kg/ha)	Yielda (kg/ha)	Multiplication ratio
25	4 081	157
50	4 907	98
75	5 176	69
100	4 949	49
125	5 574	44

Source: FAO, 1975; van Gastel and Hopkins, 1988.

In practice, very low seed rates are not used. Lower seed rates may be used when planting early generations, but certified seed is planted at the normal or slightly lower seed rate. For certified seed, acreage becomes too large to be closely monitored and thus the risk of reduced yield is too high.

Seed size

In wheat, seed size is positively correlated with seed vigour: larger seeds tend to produce more vigorous seedlings. Larger seeds of spring wheat produced higher yields than smaller seeds under latesown conditions, but not under optimum management conditions. Similarly, Khah et al. found that lowvigour spring wheat seed produced lower yields only when it resulted in low plant populations or when planting was later than normal. However, Mian and Nafziger have found that seed size has little effect on emergence of soft red winter wheat.

Fertilisation

Fertiliser application to seed crops should be based on local recommendations. A wellbalanced supply of nitrogen, phosphorus and potassium (NPK) is essential for seed production as it has an influence on seed development and thereby on seed quality. Phosphorus is essential for enhancing seed maturity and K for seed development. Ascher *et al.* concluded that seed nutrition combined with soil nutrition gave better yields and better seed quality. For example, cereals (wheat, barley and oats) grown from seed with a high P concentration gave 20 percent more yield than with low P, with a critical concentration of 0.30 percent for wheat seed. Ngugi reported that the use of compound fertilisers with sulphur (S) and calcium (Ca) gave higher grain yield and better seed quality (higher bushel weights) in non-traditional wheatgrowing areas of Kenya.

The benefit of fertiliser is not always apparent. Fertiliser may increase the disease incidence and competition of weeds, such as wild oats. High N levels may promote vegetative growth, delayed maturity, predispose the crop to foliar diseases and lead to severe lodging and reduced yield and seed quality. In Ethiopia, both N and P increased the incidence of stripe rust, though the effect of P was less pronounced. In a similar study, the density of wild oat panicles increased with N fertiliser, but decreased with P fertiliser. Top dressing wheat crops with high levels of N will make the crop lodge, which makes it difficult to be field inspected by the seed certification service.

Cleanliness of machinery

Cleanliness of machinery upon planting is very important. Seed drills should be cleaned with compressed air when changing between varieties and other crops of similar seed characteristics. Vans and trailers used for transporting the seed should be completely clean to avoid contamination.

Isolation

Isolation, growing a seed crop separate from all sources of contamination (genetic, physical and pathological), is one of the fundamental seed production techniques. In practice, such contamination can be reduced by not planting a seed crop in the vicinity of a similar crop that may contaminate it. Small, long and narrow fields need larger isolation distances than large or wide fields. Long and narrow strips are more prone to contamination than square fields. The minimum distance required for a particular crop is usually prescribed by the national seed regulations and depends on the seed class. Minimum isolation distances are larger for the early than for the later generations.

Wheat is entirely a self-pollinating crop with a very low percentage of crosspollination, from 1 to 4 percent. Therefore, the risk from genetic contamination through out-crossing is small. Appropriate isolation is, however, required to minimise physical contamination.

For wheat, it is usually sufficient to have a small strip of land between different fields to avoid mechanical admixtures. Minimum isolation distances (in metres) used in some Middle Eastern and North African countries are presented in Table 3. In India, an isolation distance of at least 3 m is required to separate a wheat seed field from the same variety not conforming to the standard.

Early generations, such as breeder seed, can be located in the middle of a field with the same variety, which is a common practice in some North African countries. Breeder seed (G1) is grown in the middle of a pre-basic (G2) field and G2 in the middle of a basic (G3) field (Plate 75) of the same variety to reduce risk of genetic and mechanical contamination.

Table 3. Examples of isolation distances for wheat in West Asia and North Africa

Country	Isolation distance (*m*)		
	Seed class		
	Pre-basic	Basic	Certified
Cyprus	—	2	2
Egypt	5	5	5
Iran	5	5	5
Sudan	4	4	4
Syria	2	2	2
Tunisia	1	1	1

Source: WANA Secretariat, 1995.

For wheat, isolation distances are more important when dealing with smut-susceptible varieties. In India, Agarwal reported that basic and certified seed fields of susceptible wheat varieties should be isolated by 150 m from fields infected with loose smut in excess of 0.1 and 0.5 percent, respectively. In Morocco, a 150 m isolation distance is required from other wheat fields if infection with loose smut is beyond 0.1 percent for pre-basic, 0.2 percent for basic, 0.3 percent for certified 1st generation and 0.5 percent for certified 2nd generation seed.

Restricting the number of varieties grown for seed multiplication per farm will reduce the chances of contamination. For example, in Tunisia only one wheat variety can be multiplied on a farm.

Crop Management

Agronomic management should be optimal and is similar to that for a grain crop. Small differences do, however, exist:

1. using lower seed rates to increase the multiplication factor;
2. leaving lanes to facilitate roguing and inspection;
3. applying slightly less than optimum amount of N to reduce lodging;
4. maintaining the species and variety purity;
5. controlling diseases that are seed transmitted.

Roguing

Roguing, removing undesirable plants, is another fundamental aspect of seed production. Undesirable plants, commonly known as rogues, are: (i) off-types or genetic variants of the same variety; (ii) other varieties of the same species; (iii) other crop species of similar growth habit and seed characteristics; (iv) noxious weeds; and (v) infected plants with seed-borne diseases. Roguing is carried out to maintain the variety and species purity of the crop and to keep the seed crop free from seed-transmitted diseases.

Roguing follows a systematic procedure and should be carried out at the time when rogues can be most easily identified and before any contamination occurs to the seed crop. The best periods for roguing a wheat seed crop are at heading and at maturity because off-types and other varieties of the same species are most easily identified. In wheat, all off-types, including those that appear due to residual segregation, should be removed. For example, the inherent problem of tall off-types in semidwarf wheat varieties due to aneuploidy, arising from chromosome pairing failures at meiosis, has been discussed by Storlie and Talbert and Worland and Law. Similarly, other crop species whose seed is difficult to separate during processing (barley, oats, triticale, rye), noxious weeds (wild oats, lolium) and infected plants with seed-borne diseases (i.e. *Fusarium, Helminthosporium, Tilletia*) are usually easily distinguished at these stages. It may be necessary to rogue the field several times, and all tillers and roots should be removed (pulled) to prevent regrowth. Rogues should be removed from the field and disposed of far away from the field or burned.

Roguing loose smut (*Ustilago tritici*) infected plants has no effect (except in reducing the disease inoculum) because the spores have already been spread and infection of the seed crop has taken place by the time the symptoms become visible. Therefore, certification schemes often do not allow the removal of smut-infected plants. However, reports from India indicated that loose smut-infected plants do head earlier and that flag leaves of smutted tillers exhibit yellow chlorotic streaks, yellow patches or yellowing at tips, even before ear emergence, and can thus be rogued before flowering. However, plants without streaks were also smutted.

Seed certification schemes set minimum standards for each class of contaminants that are permitted in a seed crop. Roguing aims at ensuring that these standards are met. Roguing should be carried out only when the seed fields do not meet these standards; roguing fields that meet the standards is not economical.

While roguing, no selection should be carried out to ensure that the genetic make-up of the variety remains the same. Therefore, roguing of early generations should preferably be carried out by the breeder or under the breeder's supervision. Roguing in early generations can be most effective.

HARVESTING

Mechanical harvesting is a common practice for seed production fields. Breeder and prebasic seed are harvested by plot combine and do not constitute many problems. Basic and certified seed, however, have to be harvested with commercial combine harvesters.

The most critical factors to be considered are seed moisture content, mechanical damage and cleanliness of equipment. For seed crops, dry weather during ripening and harvesting is essential.

Moisture

Cereal seed reaches physiological maturity between 35 to 45 percent moisture content, but it needs to dry down to a safer moisture content for harvesting and storage. The seed moisture content can be used as an indicator of when the crop is ready for harvest. Electric moisture meters or the crop characteristics can be used to decide when to harvest. For wheat, threshing or combine harvesting at 16 to 19 percent moisture content reduces mechanical damage.

Mechanical damage

Proper adjustment of the concave clearance and drum speed of a combine is essential to avoid damage to the seed crop. Mechanical damage becomes a serious problem for durum wheat seed production in Algeria and Morocco. In Algeria, seed harvested by a single ear thresher or combine showed low germination and increased fungal attack.

Cleanliness

Combine harvesters are often difficult to clean and may still harbour contaminating seed even after thorough cleaning. The availability of compressed air is important. The combine should be thoroughly cleaned before harvesting, as well as between different varieties. When harvesting the next variety, the first few hundred kilograms of seed may be discarded because contamination may still be present in the combine. For larger plots of certified seed, it is also possible to harvest and discard the outlying rows of the field.

After harvesting, the seed should be packaged in new and clean bags to avoid contamination.

PROCESSING

After a seed crop has been harvested, the seed, if necessary, has to be dried and cleaned, i.e. removal of inert matter, seed of weeds, other crops and other varieties, and seeds that are diseased, damaged and deteriorated. Cleaning can be done because wheat seeds differ in length, width, thickness, density, weight and shape.

For wheat seed cleaning, mainly screens, indented cylinders and air are used. Screens separate based on the width and thickness; a width (or diameter) separation is obtained by round screens, while for thickness separation oblong screens are used (Plate 76). Indented cylinders carry out length separation; the indents (cells or pockets) in the cylinder will, depending on their size, lift the seeds, which fit in the indents. Air separates seeds according to their behaviour in an air stream. The most important characteristic is the weight; light particles (dust, chaff, glumes or empty or partly filled seeds) will be lifted, whereas the heavier seed will fall down through the air stream. Each crop requires a different set of machines.

The raw (unprocessed) wheat seed is received from growers in bags or bulk and sampled to evaluate the need for fumigation, drying and cleaning, as well as to guide and monitor processing operations. If seed moisture is too high, the seed is first (pre-cleaned and) dried. If insects are present, the seed should be fumigated. After the raw seed has been received and where necessary fumigated, processing operations (and sequence) are as outlined below.

Pre-cleaner

Wheat seed often contains considerable plant material trash, and it is often pre-cleaned. A typical pre-cleaner is similar to an air-screen cleaner, except that it has only one air channel to remove light material, one top scalping screen to remove large particles and one bottom grading screen to remove small particles.

Dryer

If wheat seed is above 11 to 12 percent moisture, it is dried before it goes into bulk storage or processing.

Air-screen Cleaner

This is the basic cleaner, usually with two air channels and, preferably, four screens. The first air channel (head aspiration) removes dust and light materials as the seed falls from the feed hopper. The second air channel (tail aspiration) removes light seed and materials after the seed passes through the last screen. Although screen configurations vary considerably, one or two top or scalping screens remove particles larger than the good seed, and one or two bottom or grading screens remove particles smaller than the good seed. Because the average size of wheat seed varies according to the growing conditions, standard screen sizes cannot be recommended. Hand testing screens should be used to determine the exact screen perforations.

Length Separator

A length separator is almost always used to clean wheat seed. By using the proper machine configuration, shorter or longer undesirable materials (such as broken grains, weed seeds, oat, barley, etc.) are removed. Broken grains and weed seeds, which are shorter than the good seed, are removed by using cylinders with smaller indents. Larger impurities can be removed by using a cylinder with indents that lift all good seed, but contaminants (wild oats, oats or barley grains and unthreshed glumes) remain in the cylinder.

Gravity Separator

After the seed is cleaned by the air-screen cleaner and indented cylinder, it may be necessary to use a gravity separator. The gravity separator classifies a seed mixture mainly according to density or specific gravity. It can be used to remove unthreshed glumes and soil particles, which have similar sizes to wheat but different weights. Another application is the removal of weevil-infested grains from the seed lot and upgrading seed (in order to improve germination). Furthermore, wild oats and some barley may be removed from the wheat seed lots, but at the expense of substantial amounts of good seed and only after recycling the material a number of times on the gravity separator.

Treater

Wheat seed should, if necessary, be treated with the appropriate fungicide to protect the seed and seedling after planting. Insecticides are sometimes applied to protect seed in storage and in the soil. Treatments may be applied to protect the seedlings or adult plants against pathogens carried on or in the seed.

Dryer

In humid and hot climates, seeds may be sealed in vapour-tight plastic bags to maintain viability over longer periods. In such cases, wheat seed moisture content must be below 9 percent, preferably not over 8.5 percent. Usually, a dehumidified, closed circuit dryer is used after the seed treatment is applied.

Bagger-weigher

The final step is to weigh the proper amount of seed into the proper kind of bag. Wheat seed bags should be of a size that fits local farmer needs (seed rates and field size).

During processing, strict attention should be paid to the cleanliness of the processing machines and any admixture should be avoided. Every processing plant should have a complete set of hand screens, a small airscreen cleaner and an indented cylinder to help determine the proper processing requirements. It is also essential to have an internal quality control laboratory attached to each seed plant with a small seed testing facility. This laboratory unit should constantly monitor the quality of the seed and the efficiency of processing operations.

SEED TREATMENT

Seed health is an important attribute of quality, and seed used for planting should be free from pests. Seed infection may lead to low germination, reduced field establishment, severe yield loss or a total crop failure. For example, severely infected wheat grains with Karnal bunt either fail to germinate or produce a greater percentage of abnormal seedlings.

In wheat, fungi (*Fusarium* spp., *Tilletia* spp., *Drechslera* spp., *Septoria* spp. And *Ustilago* spp.), bacteria (*Corynebacterium, Pseudomonas* and *Xanthomonas*) and nematodes (*Anguina tritici*) are the most important seed-borne diseases due to their worldwide distribution and losses they incur in crop production. Chemical seed treatment is one of the efficient and economic plant protection practices and can be used to control both external and internal seed infection. It protects young seedlings or adult plants against attack from seed-borne, soil-borne or airborne pests. It disinfects seed from pathogen, checks spread of harmful organisms, promotes seedling establishment, maintains and improves seed quality or minimises yield losses. Selection of the proper chemical depends on the target organisms. A wide range of chemicals and equipment are now available for such purposes. Some recent literature gives detailed information on the management of bunts and smuts and bacterial seed-borne diseases of wheat.

Meisner *et al.* indicated that Vitavax 200 (Carboxin [37.5 percent] and Thiram [37.5 percent]) is an effective broad spectrum seed treatment fungicide, both for externally and internally seed-borne diseases of wheat. Moreover, pre-harvest foliar application of chemicals can also reduce the internally seed-borne fungi and can be combined with seed treatment to produce healthy seed. Sinclair cited that foliar spraying of wheat with benomyl, methyl benzimidazole carbamate or benomyl plus mancozeb reduced *F. graminearum*, whereas capatafol and mancozeb reduced *S. nodorum*.

Apart from disease control, seed treatment also has a positive effect on crop growth and yield. Ahmed reported that wheat seed treatment with systemic fungicides, such as Baytan, Raxil and Vitavax, significantly increased crop stand, grain yield and yield attributes. Meisner *et al.* reported a 10 percent increase in wheat yield due to seed treatment with Vitavax 200 against smuts.

Seed production in disease-free areas or under effective disease control and field inspection schemes is very important to obtain disease-free seed. Thus, understanding disease epidemiology, its transmission rate and economic threshold, combined with seed health testing, could help to define the need for seed treatment.

STORAGE

Seed is 'in storage' from the time it reaches physiological maturity on the parent plant until it is planted by the farmer. Germination is highest at physiological maturity, and viability then declines inexorably until the seed dies. Deterioration of seed viability cannot be reversed once it has occurred. Good storage cannot improve the quality of poor seed; therefore, only seed with high germination and high vigour should be put into storage. Storage conditions should then be as favourable as possible to maintain quality. Unfavourable conditions at any time during storage may reduce or destroy viability.

Seed should be harvested when it reaches harvest maturity, dried to a safe moisture content (if necessary), stored under favourable conditions and protected from damage and pests until it can be planted. Conditions that cause the loss of seed viability in storage include:

1. immature or damaged seed cannot survive long storage periods. Seed should be harvested when properly matured;
2. mechanical injury to seed during harvest or handling makes it more susceptible to deterioration in storage;
3. seed should be properly dried before going into storage and protected from moisture and high relative humidity. Fungi (*Aspergillus* and *Penicillium*) cause damage to stored seed if seed moisture is high;
4. high storage temperature has a damaging effect on seed. Stores should be designed so that low temperatures are maintained;
5. some seed treatments cause seed to die if it is stored too long; therefore, seed should only be treated when it is certain that it will be sold for planting;
6. rodents, mainly rats and mice, can be most destructive to seed. Effective rodent control (traps and poison) is essential in all seed stores. A complete programme of exclusion, sanitation and control should be used;

7. insects should be controlled by a combination of insecticides and fumigants. Use safest fumigants (e.g. phostoxin) because some fumigants (e.g. methyl bromide) will reduce germination.

Keeping the seed as dry and cool as possible in clean stores is the best management practice. If seed is dry and cool, physiological processes, fungal activity and insect activity are low. Select a seed storage site that is cool and dry (low relative humidity). O'Dowd and Dobie described the design of open seed stores, and Ellis suggested a practical guideline in choosing alternative sites for short- and medium-term seed storage.

Clements discussed the problem associated with wheat storage under tropical conditions. Wheat seed is storable for medium to long periods if kept under safe storage conditions. For wheat, high seed moisture (above 11 to 12 percent) is the most damaging, and seed must be kept as dry as possible in storage. The response of wheat seed to high atmospheric humidity (RH) in storage varies with temperature. Clements reported that at 25°C and 75 percent RH the equilibrium moisture content for wheat is 15 percent, and at 90 percent RH this may increase to 19.7%. He also stated that the critical moisture content for wheat that increases the rate of respiration is 14.6 percent. In general, stored wheat seed should be kept at moisture content levels below 12 percent and relative humidity should not exceed 50 to 60 percent.

Diekmann indicates that in cereals there are more than 20 different species of storage pests of which grain borer (*Rhyzopertha*) and weevils (*Sitophilus*) occur most frequently. Some insects, such as the khapra beetle (*Trogoderma granarium*), are quarantine pests in some countries. In India, it is reported that the rice weevil (*S. oryzae*), lesser grain borer (*R. dominica*), khapra beetle and flour beetle (*Tribolium castaneum*) are important seed storage pests of wheat (Singh, 1985). In Bangladesh, farmers traditionally use natural insecticides, such as neem (*Azadirachta indica*), biskatali (*Rumex obtusifolius*) and tobacco leaves, as insect repellents for farm-saved seed .

SEED QUALITY CONTROL

The seed quality control and certification agency is responsible for ensuring variety purity, identity and other seed quality attributes, such as physical purity, germination and health. The official seed quality control agency carries out inspection in the field and tests the seed in the laboratory to confirm that the seed meets the national certification standards.

Variety purity and identity of seed is ensured through field inspection of growing crops. Land requirement, isolation distance and seed source are confirmed, and the presence of off-types, other varieties, other crops and seed-borne diseases are determined based on inspection of representative samples that are compared with the standards. Seed crops that meet minimum field standards are accepted for harvesting and processing.

After processing, the seed is tested for purity, germination, health, etc. These tests are carried out in a seed testing laboratory based on samples taken from cleaned seed lots. After a seed crop has been field inspected and laboratory tested, and standards have been met, labels can be attached to each sealed seed bag, and the seed is officially certified.

Seed standards need to be rationalised, taking into consideration the relevance of the quality attribute, the experience of the seed growers and the level of agricultural development in the country. Excessively high standards that are copied from developed seed programmes may frustrate the growth of the seed industry and should be avoided. An example of field and seed standards for different classes of wheat seed used in Morocco is given in Table 4.

Table 4. Field and seed standards for wheat in Morocco

Standards	Seed classes			
	Pre-basic	Basic	Certified 1	Certified 2
Field standards				
Rotation (*minimum, years*)	1	1	1	1
Isolation (*minimum, m*)	10	10	10	10
Other varieties (*maximum, %*)	0.05	0.1	0.2	0.3
Other species (*maximum, number*)	1/15 000	1/10 000	1/5 000	1/2 000
Noxious weeds	—	—	—	—
Infected plants (*maximum, number*)	1/10 000	1/5 000	1/2 000	1/1 000
Seed standards				
Pure seed (*minimum, %*)	99	99	98.5	98
Other varieties (*maximum, number/1 000 seeds*)	1	1	2	3
Weed seeds (*maximum, number/kg*)	6	8	20	30
Cereal species (*maximum, number/kg*)	1	2	12	15
Wild oats (*maximum, number/kg*)	0	0	1	1
Noxious weeds (*maximum, number/kg*)	3	4	6	8
Infected seeds				
Loose smut and common bunt (*maximum, %*)	—	0.02	0.05	0.1
Insect infestation				
Live insects (*maximum, number*)	0	0	0	0
Insect damage (*maximum, %*)	0.1	0.1	0.1	0.1
Germination (*minimum, %*)	85	85	85	85
Moisture content (*maximum, %*)	14	14	14	14
Specific gravity (*minimum, hl/kg*)	79	79	79	79

Source: WANA Secretariat, 1995.

In many developed countries, there is a tendency to give more importance to the internal quality control by the companies that are producing and marketing the seed. In developing countries, seed quality control agencies often lack resources and are not able to carry out proper seed quality assessment. In view of this, the Food and Agriculture Organisation of the United Nations (FAO) proposal *Quality Declared Seed* is particularly relevant to developing countries because it aims at a system that requires fewer resources and still provides sufficient guarantee for the farmers.

MARKETING

Seed marketing is the final step in a seed programme; it takes the seed to farmers and gets them to buy it and plant it. Seed marketing is time-sensitive and sensitive to factors affecting rural marketing. Seed must reach the farmer at the right time, at the right place, at the right price, in the right amount and must be of the highest quality. Because seed marketing is sensitive to many factors, it is often considered a high-risk business. Seed marketing requires:

1. convincing farmers that the seed quality is high and ensuring that only high-quality seed is sold;
2. convincing farmers that the seed quality means a benefit to them that is worth the extra cost they must pay for the seed. All possible means of promoting the seed must be used;
3. making the seed available in locations close to the target farmers. The seed must be readily available when the farmer needs it. It must be available in bag sizes that fit the farmers' needs.

Effective seed marketing is a specialised operation, as described by Gregg and van Gastel. Wheat seed marketing has the same requirements, problems and needs as the marketing of other crop seed. Farmers' variety and seed replacement is influenced by their perception of the yield gain of a new variety, the yield loss of an old variety and the risk in changing the variety. In wheat, decline in variety purity and yield from retained seed is expected to be low. An estimated yield reduction of 0 percent in dry areas of Australia, 0.25 percent in Pakistan and 1.6 percent in Nepal from retained seed is reported. As a result, farmers can save their own seed for planting for quite a long period of time once they have purchased seed of the new variety. Few farmers are willing to spend more money to purchase improved seed of wheat. On average, wheat farmers typically purchase a relatively small amount of improved seed.

REFERENCES

Agrawal, V.K. & Gupta, M.C. 2019. Loose smut of wheat - A treat to seed production. *Seed Res.*, 17(1): 55-68.

Boyd, A.H., Dougherty, G.M., Matthews, R.K. & Rushing, K.W. 2015. Seed drying and processing. *In* W.P. Feistritzer, ed. *Cereal seed technology: a manual cereal seed production, quality control, and distribution*. Rome, FAO.

Clements, D.J. 2017. Wheat seed storage under tropical conditions. *In* A.R. Klatt, ed. *Wheat production constraints tropical environments*, p. 360-365. Mexico, DF, CIMMYT.

Diekmann, M. 2021. Seed-borne diseases seed production. Aleppo, Syria, ICARDA.

Ellis, R.H. 2018. The viability equation, seed viability nomographs, and practical advice on seed storage. *Seed Sci. Tech.,* 16: 29-50.

Laverack, G.K. 2014. Management of breeders seed production. *Seed Sci. Tech.,* 22: 551-563.

Mamluk, O.F. & van Leur, J. 2016. Transmission of seed-borne diseases of wheat and barley. *In* J.P. Srivastava & T.L. Simarski, eds. *Seed production technology.* Aleppo, Syria, ICARDA.

Wilcoxson, R.D. & Saari, E.E., eds. 2016. *Bunt and smut diseases of wheat:* concepts and methods of disease management. Mexico, DF, CIMMYT.

Worland, A.J. & Law, C.N. 2015. Aneupolidy in semidwarf wheat varieties. *Euphytica,* 34: 317-327.

3

Hybrid Rice Seed Production

Commercial exploitation of hybrid vigor is one of the most important applications of genetics in agriculture. It has not only contributed to food security, but has also benefited the environment. It is the basis of a billion-dollar agribusiness which has generated significant employment opportunities all over the world. Among the various crop species in which hybrid varieties are used commercially, rice ranks very high. Rice hybrids were first commercialised in the late 1970s in China. During the past decade, Vietnam, India, the Philippines, Bangladesh and the United States have also begun the commercial production of hybrid rice.

About 15 million ha (50%) of rice in China are hybrid varieties, producing 103.5 million mt of paddy annually (an average yield of 6.9 mt/ha). The remaining 50%, planted in inbred high-yielding varieties, produces 81 million mt (5.4 mt/ha). Thus, on average hybrid rices in China yield about 27% (1.5 mt/ha) more than the inbred high-yielding varieties.

The commercialisation of hybrid rice technology in tropical countries has been facilitated by the research initiated in 1979 at the International Rice Research Institute (IRRI). It was soon recognised that Chinese rice hybrids were not suited to the tropics, and their grain quality was poor compared to the popular inbred high-yielding varieties grown in tropical countries. Therefore, IRRI researchers concentrated on developing suitable parental lines specially for the tropics, using the CMS system which had been found to be effective in China.

Within a decade, some commercially usable CMS and restorer lines and some elite hybrids were identified and shared with NARS for evaluation and utilisation. Some of these hybrids were released for commercial cultivation in the Philippines (IR64616H, IR68284H), India (IR64611H, IR64618H, IR69690H), Vietnam (IR64615H, IR64616H, IR69690H) in the years 1993-1998. Some IRRI CMS lines, IR58025A and IR62829A, were also used by national programs to develop local hybrids in India (e.g. APHR-1, APHR-2, CORH 2, KRH 2, ADTRH 1, etc) and the Philippines. Concurrently, several private seed companies in Asia were also provided with IRRI-bred parental lines.

Currently, about 610,000 ha are covered with hybrid rices in Vietnam (340,000 ha), India (200,000 ha), Bangladesh (50,000 ha), USA (10,000 ha), Philippines (5,000 ha) and Myanmar (5,000 ha). These have given on average about 20-25% (1 - 1.5/ha) higher yields than the inbred HYVs, thereby contributing towards higher on-farm productivity. Currently, hybrid rice technology is considered a viable option to increase rice yields globally.

Since rice is a self-pollinated crop, hybrid seed production must be based on male sterility systems. These may be cytoplasmic (CMS), thermo-sensitive genic (TGMS) or photo-sensitive genic (PGMS). Most of the commercial hybrids used in China and elsewhere are based on a CMS system, although TGMS and PGMS systems are also becoming used commercially (Yuan, pers. comm.). Hybrid seed production technology involving these male sterility systems is complex and labor-intensive. It also requires a sophisticated seed industry infrastructure.

WHAT IS HYBRID RICE?

Over the past four decades, technological innovations in rice breeding became a frontier for increasing rice productivity in Asia. Dubbed as the Green Revolution, the development and massive promotion of the so-called "modern" or "high-yielding varieties" with combined package of technologies such as synthetic fertilizers, pesticides, mechanized land preparation and irrigation transformed the landscape in most rice farming communities in Asia. Rice yields initially increased.

However, the Green Revolution suffered major setbacks when the yield of modern varieties plateaud in the 1980s. Depletion of soil fertility, negative impacts of synthetic pesticides and fertilizers to the environment and the human health, economic dislocation in rural communities, increased dependence of farmers to agro-chemical inputs and genetic erosion are just some of the issues and impacts of the Green Revolution.

While modern varieties were spreading across Asia in the 1970s, scientists in China have successfully developed the hybrid rice technology that was able to outyield the modern varieties. With subsequent massive promotion of the technology, hybrid rice was attributed as the factor why the country became rice self-sufficient and was regarded as responsible for feeding its more than a billion population.

Hybrid rice is the first generation (F1) crop grown from the cross of two distantly related rice varieties. Hybrid rice exploits the phenomenon called *hybrid vigor* or *heterosis* where the hybrid rice plant is more superior to any trait of its parents. Due to hybrid vigor, hybrid rice has 15-30% yield advantage over the conventional rice that farmers grow.

Figure 1. Rice seeds

Rice is innately a self-pollinating plant which means that each seed is self-fertilized. This enabled farmers to continually save, re-use and exchange harvested seeds through generations without significant change in character or yield performance. Traditional varieties, landraces, farmers' varieties, and even the modern varieties are referred to as *selfpollinating rice* or *inbreds.*

The development of hybrid rice requires the introduction of male sterility system that would enable the transfer of pollen from one variety to a distantly related variety to produce the F1 hybrid seeds. Unlike in the inbreds, if the harvested seeds from F1 hybrid plants are replanted, the resulting plants will be non-uniform and unstable in character causing significant yield losses for farmers.

The most common male sterility system used in national hybrid rice programs across Asia is the cyto-genetic male sterility system (CMS). In this system, the sterile female parent called A-line (also called CMS-line) is crossed with the fertile male parent called the restorer or Rline to produce the fertile F1 hybrid seeds. Through the process of breeding, a maintainer or B-line is used to multiply the sterile A-line.

PROMISES OF HYBRID RICE

Other than the 15-30% yield advantage over inbreds, scientists justified that hybrid rice is the only technological option currently available to address the dwindling rice production because:

1. Yield levels of modern varieties developed during the Green Revolution era have plateaud;
2. Increasing demand for rice especially in developing countries;
3. Hybrid rice has also shown better performance under adverse conditions like drought and saline conditions.

China remains to be the country with the widest scale of hybrid rice cultivation across Asia covering about 15 million hectares. Outside of China, there is increasing hybrid rice cultivation in Vietnam, Bangladesh and the Philippines during the recent years (Table 1).

Table 1. Areas of Hybrid Rice Cultivation in Asia

Countries	Country Hectarage of adoption (hectares)
China	15,215,000
India	280,000
Bangladesh	40,000
Vietnam	480,000
Philippines	208,342
Indonesia	20,000
Myanmar	50,000

Promotion of hybrid rice in China was heavily subsidized by the Government. Farmers were provided with free seeds, supported by subsidies on fertilizers and pesticides and were instructed to plant hybrid rice. There was a ready market for hybrid rice as hybrid rice seed and grain procurement were done by the state. There was massive adoption of hybrid rice in China such that in 1992, more than 50% of its irrigated rice areas were planted to hybrid rice.

Inspired by the success of China, IRRIc in collaboration with the Peoples' Republic of China, APSAd, FAOe and NARSf of several Asian countriesg initiated a project to promote hybrid rice technology outside of Asia. These led to the commercial release of hybrid rice varieties in the early 1990s and were subsequently promoted as national hybrid rice programs. These efforts received various financial and technical support projects from both FAO and ADB. IRRI became the main source of parental lines of hybrid rice to NARS and even to the private sector. APSA facilitated the entry of private sector participation in hybrid rice dissemination and complementation of private seed interests to national hybrid rice programs.

HYBRID RICE SEED PRODUCTION TECHNOLOGY

Hybrid rice seed production technology is different from that for inbred rice seed production, and is more complex than the seed production for many other hybrid crops. Currently, the most popular male sterility system is the CMS (popularly known in China as the three-line system). This utilises three different lines, namely a cytoplasmic male sterile line (A line), a maintainer (B line), and a restorer (R line). Using this system for the commercial production of hybrid seeds involves two major steps. The multiplication of the A line (female parent) must be carried out before the production of hybrid (F_1) seed.

The other system, which is gaining popularity in China and Vietnam, involves a seed parent which is an environmentally sensitive genetic male sterile (EGMS) or S line. This system requires neither a maintainer (B) nor a restorer (R) line. The seed parent can be multiplied just like any conventional inbred when planted in a fertility-inducing environment. When planted in a sterility-inducing environment, it becomes male sterile and may serve as a seed parent in producing F_1 seed of a hybrid. Any line found to form heterotic combinations with a particular S-line can be used as its pollen parent to produce a hybrid. Since this system involve two lines (TGMS or PGMS, and a pollen parent) to produce the hybrid seed, this system is also known as a two-line system. The three- and two-line systems for hybrid rice seed production.

HYBRID SEED PRODUCTION MECHANISM IN RICE

For hybrid rice seed production to succeed, a sufficient number of pollen grains must be deposited on the stigma lobes of each spikelet of the seed (male sterile) parent. It helps if the pollen parent grows to a greater height than the seed parent. Other plant characteristics which influence this include small and horizontal flag leaves, the number of panicles per square meter, the number of spikelets per panicle, good panicle exsertion, and synchronised flowering of seed and pollen parents. The floral traits influencing outcrossing in rice include: stigma size, style length, stigma exsertion (in seed parent) and anther length, filament length, and pollen number/anther (in pollen parent). The flowering behavior traits influencing outcrossing in rice are: number of days of blooming, time of blooming, duration of blooming, duration of floret opening, and angle of floret opening. The environmental factors influencing outcrossing in rice include temperature, relative humidity, light intensity and wind velocity. The relationship of all these characteristics is discussed at length by Virmani. In China, conditions favorable for good outcrossing in rice have been identified as: a daily temperature of 24-28°C, a relative humidity of 70-80%, a diuranal difference in temperature 8-10°C and sunny days with a breeze. Suitable field conditions include: fertile soil, a dependable irrigation and drainage system and a low risk of disease and insect infestations.

HYBRID RICE PROMOTION

Over the past decade, hybrid rice promotion has been beset with problems that hamper continuous adoption by rice farmers:

Inconsistent Yield Advantage over Inbreds

Contrary to claims that it poses 20-30% yield advantage over conventional inbred rice varieties, hybrid rice performance is location-specific and is ideal in sub-tropical and subtemperate zones. Experiences in some Asian countries attest the fact:

1. China – only 15-16% yield advantage over inbred counterparts in Jiangsu Province in 1984. Yield of hybrid rice varieties developed from CMS systems are reportedly declining;

2. Vietnam – hybrid rice cultivation is concentrated in the Northern Vietnam and during the winter months;
3. Indonesia – yield advantage of 2 hybrid rice varieties, Maro and Rokan were reportedly inconsistent in the multi-location and multi-season yield trials;
4. Philippines – IRRI scientist claims that maximum yield advantage is just 10% over the best inbred rice varieties. In a collaborative study done by SEARICE, yield advantage of 20-30% was achieved in only 1 out of the 4 top hybrid rice provinces.

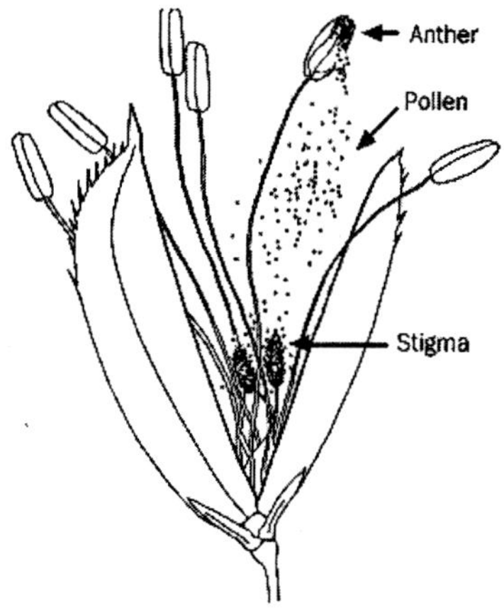

Figure 2. Inbred rice seed

Hybrid Seeds are Vulnerable to Seed-borne Diseases

Hybrid rice seeds, as compared to inbred rice seeds, are partially opened, discolored, halffilled and lightweight. Being partially opened, it is more susceptible to seed borne insect pests and diseases especially under humid tropical conditions.

In the Philippines where such conditions prevail, hybrid rice seeds need cold dry storage facilities which proved to be more costly for seed growers to maintain. In China and Northwestern India, CMS lines in hybrid rice seed production plots were reportedly found to have a higher incidence of seed-borne diseases i.e. puddy bunt and Tak and False smut.

Inferior Grain Quality

Despite long slender grains, there is a lower price for hybrid rice grains than premium inbred varieties in the market. Lower market price for hybrid rice grains in China, Vietnam, India and Bangladesh were reportedly due to its inferior quality. In Thailand, the lack of demand for hybrid rice is mainly due to market preference for Jasmine Rice;In the Philippines, despite higher premium of hybrid rice compared with ordinary rice, milled hybrid rice is sold as ordinary rice because of its low head rice recovery.

Costly Seeds and Technology

Hybrid rice seed production is labor, input and knowledge intensive. It requires additional 50 man-days/ha than normal mechanized rice cultivation. It is highly dependent on the use of gibberellic acid, a growth regulator needed to synchronize the flowering of the hybrid seed parents. Outside of China, giberellic acid is expensive (more than $1.00 per gram).

1. Philippines – regular unsubsidized cost of hybrid rice is PhP 2,400 per hectare (US$48) compared to certified inbred variety which on the average costs PhP 1,400 (US$28) per hectare;
2. India - average cost is Rs 1500 per hectare while inbred varieties cost Rs 300-400;
3. Vietnam – hybrid rice is reportedly sold at US$1.21 per kg as compared to US0.22 per kg for inbreds.

Hybrid rice cultivation requires more fertilizers than ordinary inbred rice production as observed in India, Vietnam and the Philippines.

Irrigation and transplanting of rice seedlings warrant successful hybrid rice cultivation. In China, almost 100% of the rice area is irrigated and transplanting is practiced in 97% of the rice area. Dramatic change of cultivation and farming systems is required if hybrid rice will be planted in rainfed and upland conditions.

Lower Net Income for Farmers

Promise of yield increases did not necessarily translate into net profit for farmers planting hybrid rice:

1. *Philippines:* insignificant difference in net income between hybrid rice and inbred rice varieties despite findings of higher yield in 3 out of 5 provinces due to higher seed, fertilizer, pesticide and hired labor cost;
2. *India:* experience in Karnataka showed hybrid rice production posed lower profit (net return) due to additional costs on inputs.

RISKS POSED BY HYBRID RICE TO FOOD SOVEREIGNTY

Loss of genetic diversity has become a major concern in modern varieties in Asia where varieties have become more vulnerable to pests and diseases. Around 91% of the post IR-8 varieties are genetically related and have a single parental source which can be traced to a single cytoplasm called *Cina*.

With hybrid rice, problem of cytoplasmic uniformity is graver where 95% of the total area planted to hybrid in Asia came from a single cytoplasmh source – the WA cytoplasm utilized in the CMS system. Albeit efforts to diversify the CMS lines in rice, about 88% of the commercial hybrids in China carry WA cytoplasm. This cytoplasmic uniformity could result in disease epidemics similar to the Southern corn blight outbreak that wiped out US hybrid maize due to the use of a single cytoplasm.

Curtailment of Farmers' Rights to Save and Re-use Seeds

Hybrid rice is only economically viable if planted for one cropping season. If hybrid rice is used as seeds during the subsequent cropping season, resulting plants will be nonuniform and yield will deteriorate. Hence, farmers are technologically restricted of any form of seed saving and would have to buy new seeds every cropping season. Farmers are further restricted to freely use hybrid rice in countries that have IPRi legislations on seeds like PVP j laws and/or PBRsk. Farmers are legally prohibited from creating new strains from hybrid rice varieties or from selling new strains due to provisions against use of essentially derived varieties.

This displaces the dynamic system that sustained rice farming in many communities for centuries. Farmers' traditional rights to save, re-use and exchange seeds, already recognized in the International Treaty on Plant Genetic Resources in Food and Agriculture (ITPGRFA) are not recognized in hybrid rice systems.

In Vietnam, a study in 1999 warned that hybrid rice poses a great logistical challenge for the seed supply systems to annually renew all seed. A crop failure in the seed multiplication system can be disastrous for farmers as farmers cannot rely on harvested grains for seeds. As it is, 35-45% of the total cultivated areas of Vietnam rely on farmsaved seeds and 8-18% rely on exchanged seeds.

Possible Displacement of the role of women-farmers

In Vietnam, the women play a key role in all activities involving farm-seed saving i.e. handling, harvesting to raising seedlings for the succeeding cropping season. However, their role will be displaced if Vietnam will continue to promote hybrid rice on a massive scale.

Opening the Flood Gate to Corporate Control of the Rice Seed Sector

Hybrid rice is setting the stage of the 'privatization' of the rice seed sector despite the key role played by public research institutions in the initial stages of hybrid rice programs. Already, there is an increasing trend of private seed companies involved in hybrid rice promotions across Asia (Table 2).

Table 2. Some Hybrid Rice Seed Companies in Asia.

Company	Subsidiaries and joint ventures	Country/ies Present	Hybrid Rice Varieties
Bayer (Germany)	HRI Ltd, Hyderabad	India Philippines	PA 6201, 6444 Bigante
Pioneer Seed Co.	SPIC PHI Seeds Ltd.	India	PHB-71
Monsanto (US)	Mahyco	India	Suruchi 5170,5319,5445; white jewel 1,2,3,4
	Cargill	Philippines	Magilas 500
Yuan Longping High Tech Agriculture (China)	Guard Rice PT Bangun Pusaka Aftab Bahumukhi Farm/Islam Group Rice Tec (USA)	Bangladesh Indonesia	
SL Agritech (Philippines)		Philippines	SL-8H
East-West Seeds	Hyrice Corporation	Philippines	Rizalina 20
RB Biotech	Chua Family Foundation, Yayasan Tuanku Syed Sirajuddin, Sunland Agritech and West Wishes Agritech	Malaysia	
Bioseed	Bioseed Shriram Bioseed Ltd, Hyderabad	Philippines India	MRH005

Privatization of the rice seed industry means that farmers who want to use hybrid rice seeds will have to depend on the seed access, supply, terms and conditions set by seed companies.

GUIDELINES FOR HYBRID RICE SEED PRODUCTION

Seed yield from a male sterile line used in hybrid rice seed production is a function of three factors. First is the yielding ability of the male sterile line, as determined by the yielding ability of its maintainer, the fertile counterpart. Second is the proportion of the male sterile line to the pollen parent. The third is the outcrossing rate of the male sterile line. Improvement in any of these three factors can help to increase the seed yields of hybrid rice.

Extensive research has led to the identification of the following guidelines for successful hybrid rice seed production.

1. Selection of seed and pollen parents with synchronised time of anthesis.
2. Selection of seed parents with long, exserted stigma, longer duration, and wider angle of floret opening.
3. Selection of a pollen parent with a high percentage of residual pollen

per anther after anther exsertion. High pollen shedding potential is attained by getting 2000 - 3000 spikelets/m^2 to bloom per hour during peak flowering period.

4. Synchronisation of flowering time of the two parents by seeding them at different dates depending on their growth duration or estimated accumulated temperature requirements for initiatiion of flowering.
5. Use of optimum seed parent:pollen parent row ratio such that the ratio of spikelet number per unit area of seed parent and pollen parent is about 3.5:1.
6. Use of seed and pollen parents with small and horizontal flag leaves, or cutting long and erect flag leaves.
7. Use of gibberellic acid (GA_3) to improve panicle exsertion and prolong duration of floret opening and stigma receptivity.
8. Planting of seed parent pollen parent rows across the prevailing wind direction and use of supplementary pollination with a rope or stick when wind velocity is below 2.5 m/sec.
9. Selection of optimum time of flowering of parential lines in seed production plots.

PRACTICES FOR HYBRID RICE SEED PRODUCTION

Practices involved in producing hybrid rice seeds using a male sterility system include the following.

Use an Isolated Field

A field should be used which is isolated from other rice varieties by distance (50-100 m), time of flowering (at least 3 weeks) or some natural or artificial barrier. The isolation distance can be decreased by increasing the number of pollen parent rows around the seed production plot. Barrier isolation can be effected through topographical features, or by the use of tall plants or artificial obstacles to block pollen movement.

Timing of seeding of parent lines

The two parental lines should be sown and transplanted at the right times so that their flowering is synchronised. This is the most critical step in hybrid seed production, and is practiced by using one of the three methods described below:

1. The seeding date of the two parents can be determined according to the difference in the time they take to reach maturity. The parent which matures earlier must be sown later than the other parent.

2. The seeding date of the two parents can be determined by the difference in the number of leaves at the main culm. If this trait is stable, the date of seeding of the early heading parent can be decided according to the predetermined number of leaves present on the later heading parent.
3. The differential seeding of parental lines can be based on their effective accumulated temperature (EAT) from seeding to flowering. The EAT of a particular variety is calculated by summing up the mean temperature of days with a range within 12°C to 27°C. The parent which has a higher EAT is sown earlier, and the other parent is sown after the difference in EAT between the two has accumulated.

Optimum Ratio of Seed Parent: Pollen Parent Plants

The optimum seed parent:pollen parent ratio must be used, according to the spikelet number per panicle of parental lines, and the size of the pollen load of the pollen parent. The seed parent is planted densely, to increase the number of spikelets per square meter. The pollen parent is more widely spaced, to promote tillering and help ensure that the pollen load is greater over a longer period.

The optimum ratio for plants of the seed: pollen parent should ensure 3.5 spikelets of the seed parent for each spikelet of the pollen parent. Some commonly used seed parent:pollen parent ratios are 10:3, 6:2 and 14:2. In extreme cases, a ratio of 16:2 has been used. In the United States, a ratio of 18:6 is used for mechanised hybrid seed production.

Good Cultural Management

The seed production plot must be given good cultural management. Weeds, diseases and insects must be controlled, and the optimum fertiliser rate must be applied. If the panicle initiation stage of the two parental lines is found to occur at different times, they should be given different application rates of nitrogen, phosphate and potassium fertiliser, to adjust the flowering date of the parental lines by 2-5 days.

If the parental lines have long, erect flag leaves which might obstruct the movement of pollen, the flag leaves should be clipped by one-half or two-thirds of their length, one to two days before the initial heading.

However, this practice has the negative effect that it may spread diseases such as bacterial blight, and sheath blight through the wounds inflicted on the leaves. Besides, this practice also reduces photosynthetic activity. Parental lines possessing short narrow, flag leaves do not need to be clipped.

GA_3 should be sprayed on parental lines which show poor panicle exsertion. Besides improving panicle exsertion, GA_3 also increases the stigma exsertion rate, as well as the time during which the spikelet is open and the stigma is receptive. Differential application of GA_3 also increases the relative height of the pollen parent over the seed parent. GA_3 is generally sprayed using a knapsack sprayer. Spraying begins when 5-10% of the rice plants are at the heading stage, and lasts for 2-4 consecutive days. Spraying should be done on a calm sunny day, usually in the afternoon. It should not be carried out if rain is expected within 24 hours.

Hybrid rice seed producers in China use a very high dosage (150-300 gm/ha) to get high seed yields. However, outside China the high cost of GA_3 limits seed growers to using only 45-50 gm/ha. The use of an Ultra Low Volume (ULV) sprayer has been found to reduce GA_3 dosage to 15-20 gm/ha, and give seed yields similar to those from spraying 45-50 gm GA_3 from a knapsack sprayer.

The horizontal dispersal of pollen grains can be increased by shaking the panicles of the pollen parent by pulling a rope between two people, or stirring the canopy layer of the pollen parent with a bamboo stick (supplementary pollination). This should be done during the anthesis of the pollen parent. Either method may increase outcrossing, and hence seed yields. Care should be taken that panicles are not broken at the neck.

Roguing of off-types should be carried out at maximum tillering, flowering and maturity stages, to prevent any cross-pollination between them and the true-to-type seed and pollen parent plants.

Pollen and seed parents should be harvested, threshed, dried and cleaned separately and carefully, to avoid any mechanical mixture of seed of parental lines. Using the above practices, seed yields of 0.5 to 4 mt/ha (mean 1.3 mt/ha) have been obtained in the tropics. However, because of the vast experience of China's seed growers and specialised hybrid rice seed production, yields in China range from 1.5 mt/ha to 7.4 mt/ha (mean 2.5 mt/ha). In the late 1970s, average hybrid rice seed yield in Mainland China were 0.67 mt/ha. Currently, these have increased to 2.5 mt/ha. This increase in seed yield is a associated with the modified practices shown in Table 3.

QUALITY CONTROL FOR HYBRID RICE SEED

Quality control is vital to the widespread adoption of hybrid rice. Farmers who buy expensive hybrid seeds expect high quality. Quality control is carried out throughout the whole process of seed production. It includes the nucleus, breeder, foundation and certified seed production of both parental lines and F_1 hybrids.

Seed quality standards are met by intensive roguing of off-types (from seeding to harvesting) and careful handling during harvesting, threshing, drying, cleaning, processing, bagging and labeling. Grow-out tests, germination tests, seed vigor tests etc. should also be done before the seeds are sold commercially. China has a well-established hybrid rice seed industry and national standards for hybrid rice. Seed quality standards are in place (Table 4) and being followed. In other countries where hybrid seed production is more recent, standards vary and are still to be formalised.

Table 3. Progressive modification of techniques of hybrid rice production in China during the late 1970 to late 1990s to increase seed yields.

Technique	Period		
	Late 1970s	Late 1980s	Late 1990s
Row ratio	1-2:6-8	2:10-12	2:14-18
Width of a female parent strip (m)	1.5	2.0	2.5
Plants/hill			
A line	1	2	2
R line	1	1	3-4
Transplanting density (cm)			
A line	16.5 × 16.5	15 × 15	13 × 13
R line	20 × 20	20 × 20	17 × 30
Hills/ha (× 1000)			
A line	320	450	590
R line	67	50	27
Plants/ha (× 1000)			
A line	320	900	1060
R line	67	50	100
Plants/ha (× 1000)	2100	2600	3200
Dosage of GA_3 (g/ha)	0-45	75-120	180-220
Concentration (ppm)	20-40	40-60	100
Seed set percentage	10-15	25-30	45-50
Clipping of flag leaves	Yes	Yes or no	No or lightly
Yield (mt/ha)	0.54	1.80	2.50

VEGETATIVE PROPAGATION OF HYBRID RICE

Vegetative propagation of rice is wellknown. Adult plants can be raised from seedlings, tillers, culm cuttings or by ratooning. Clonal propagation of rice was proposed in early sixties to exploit hybrid vigor in rice. However, no serious attempt was made to develop this as a technology. Since the commercialisation of hybrid rice technology in China in 1976, agronomists and farmers have tried using vegetative propagation in various ways to reduce the seed rate, and hence the seed cost, of commercial rice hybrids. These approaches include double transplanting and ratooning.

In double transplanting, hybrid seeds are sown densely in the seed bed. When the seedlings reach the 2-3 leaf stage, they are transplanted into a field nursery, using a wide spacing of 33 × 10 cm to induce more tillering. After 25-30 days, tillers are separated and transplanted into the main rice field. Although double transplanting can reduce the seed rate of hybrid rice significantly, it does increase the labor cost. In subsistence farming, where family labor is readily available and there is no other opportunity cost, this method may be useful. However, in situations where labor is either not available or can earn more elsewhere, this approach may not be economical.

Table 4. Seed standards for parental lines and F_1s of hybrid rice in China

Seed grade		Purity (>%)	Cleanness (>%)	Germination (>%)	Moisture (>%)	Weed seeds (> grains/kg)
A line	Nuclear seed	100.0	99.8	98.0	3.0	0
	Foundation seed	99.9	99.	96.0	13.0	0
	1st class seed	99.5	99.0	96.0	13.0	0
	2nd class seed	99.0	97.0	95.0	13.0	5
B line	Nuclear seed	100.0	99.8	98.0	13.0	0
	Foundation seed	99.9	99.0	96.0	13.0	0
	1st class seed	99.5	99.0	96.0	13.0	0
	2nd class seed	99.0	97.0	93.0	13.0	5
R line	Nuclear seed	100.0	99.8	98.0	13.0	0
	Foundation seed	99.8	99.0	96.0	13.0	0
	1st class seed	99.5	99.0	96.0	13.0	0
	2nd class seed	99.0	97.0	93.0	13.0	5
F_1 hybrid	1st class seed	98.0	98.0	93.0	13.0	0
	2nd class seed	96.0	97.0	90.0	13.0	5

Note: 1. The 1st and 2nd class seeds are certified seeds.
2. The sterility and sterile plants of nuclear and foundation seeds of A line both should be 100%.
3. The restoring rate of nuclear and foundation seeds of R line both should be higher than 85%.

Ratooning of hybrid rice has been tried with varying degrees of success. Recently, successful vegetative propagatioin of rice has been achieved by splitting and transplanting the ratoon seedlings of an F_1 rice hybrid in Bangladesh (M.A. Mazid, pers. comm..). This approach would be practical with resource poor farmers who can use family labor but who cannot afford the full cost of hybrid seed.

ECONOMIC VIABILITY OF THE PRODUCTION OF HYBRID RICE SEED

The availability of quality seed at a reasonable price is a crucial factor behind the large-scale adoption of hybrid technology for any crop. This has been demonstrated by the success of hybrid varieties of maize, pearl millet, sorghum, sunflower and cotton, all of which are supported by economical and efficient seed production and distribution systems in the countries where the hybrids are commercialised. Chinese success in the largescale adoption of hybrid rice can also be attributed to the efficient and economical hybrid seed production and distribution system, organised by the state-owned seed industry.

Hybrid rice seed production technology is considered labor- and knowledge-intensive. It involves various risks, especially in the early stages when seed producers are still lacking in experience. Typical problems are poor synchronisation of the parental lines, and unfavorable weather. At present, commercial production of hybrid rice seed outside China is limited to India, Vietnam, the United States and the Philippines. Seed growers will engage in hybrid rice seed production only if it is more profitable than inbred varieties, and if the additional profit is enough to compensate for the greater risks and level of skill involved.

India

Studies conducted in the Indian states of Andhra Pradesh, Karnataka and West Bengal showed that on average, labor accounted for about 48% of the total cost of hybrid rice seed. Hybrid rice seed production requires about 340 mandays/ha, about 50 mandays/ha more than for ordinary rice cultivation. Hybrid seed production requires additional labor for extra farm operations such as thin and row planting, supplementary pollination, filling gaps, roguing, GA_3 application, and manual harvesting, threshing and cleaning.

China

In China, the production of hybrid rice seed takes about 375 mandays/ha, 75 mandays/ha more than ordinary rice.

The cost of female parental line seed (A seed) and gibberellic acid (GA_3) (additional inputs for hybrid seed production) accounted for 9% of the total cost (Table 5). The average cost of producing hybrid seed was US$0.52/kg. The private sector was able to buy hybrid seed from seed growers at US$0.85/kg, while government agencies paid US$1.0-1.1/kg. At an average price of US$1.0/kg, with an average seed yield of 1250 kg/ha, hybrid rice seed production was 65% more profitable than ordinary rice cultivation (Table 5).

Seed companies in India sell hybrid seed produced in India to farmers at US$2.0-2.5/kg. This large margin is needed to keep the private seed companies in business, since the demand for hybrid rice seed is small and the market is dispersed. However, as the technology picks up, economies of scale should reduce the cost of production and increase seed yields. This would allow seed companies to sell hybrid seed at a lower price to farmers. Greater competition in the seed business will also force seed companies to reduce the selling price of hybrid rice. As reported in Table 6, the selling price of hybrid rice seed could be reduced from US$2.50 to 1.0/kg, if seed yields were to increase from their present level of 1.25 mt/ha to 3.0 mt/ha. In such a case, the cost of hybrid seed would fall by US$20-30/ha.

Table 5. Comparative cost-return profile for hybrid rice seed production and commercial inbred rice production in India (1997-98 crop year)

Hybrid seed	Inbred rice production (US$/ha)	Costand returns production (US$/ha)
Cost		
Seed[a]	70.0	14.0
Fertilisers & manure	123.8	86.8
Gibberellic acid (GA_3)	50.0	—
Labor	315.0	112.5
Plant protection	15.0	13.8
Miscellaneous	71.2	11.9
Total input cost return	645.1	239.0
Yieldb (kg/ha)	1250	5950
Market pricec (US/kg)	1.0	0.1173
Return to hybrid seed	1250	—
Return to by-product (male seed and straw)	212.5	41.1
Gross return	1462.5	739.0
Net return	817.4	500.0
Unit cost of production (US$/kg)	0.516	0.040

Source: Janaiah and Hossain 2000

Table 6. Effects of increase in seed yield of hybrid rice on cost of seed production and selling price of seed in India, a sensitivity analysis

Scenario	Cost of production (US$/kg)	Selling price (US$/kg)	Seed cost for commercial cultivation @ 20 kg/ha (US$/ha)
I. At current seed yield of 1,250 kg/ha	0.52	2.50	50
II. At seed yield of 2,000 kg/ha	0.32	1.60	32
III. At seed yield of 2,500 kg/ha	0.26	1.30	26
IV. At seed yield of 3,000 kg/ha	0.22	1.00	20

Vietnam

Vietnam, because of its socio-economic and political structure, has been able to develop a strong research-extension-farmer linkage in seed production and distribution, similar to that of China. The top priority for Vietnam's hybrid rice program is to develop an efficient system for the production and distribution of hybrid seed at affordable prices. At present, about 20% of hybrid rice seed requirements are met by seed produced in vietnam, while the remaining 80% is imported from China. About 600 ha in Vietnam was under hybrid rice seed production in the year 2000, compared to only 173 ha in 1992. Average seed yields increased from about 300 kg/ha to 2500 kg/ha during this period. According to Vietnam's Ministry of Agriculture and Rural Development, hybrid rice seed requirements are about 24,000 mt, if hybrid rice cultivation is to be increased to 800,000 ha by 2005.

The total cost of hybrid rice seed production in Vietnam for the spring crop in 2000 was about US$1200/ha, of which the government provided 30%

as a subsidy in the form of free parental line seeds, GA3, chemical fertilisers, etc. Therefore, the cash cost of seed production paid by the growers was US$800/ha (Table 7). Average seed yields are 2.0 mt/ha, and the current seed procurement price is US$0.60/kg. Thus, the production of hybrid rice seed in Vietnam generates about US$400/ha net profit.

Table 7. Cost-return profile for hybrid rice seed production in Vietnam, 2000, spring

Particulars	Values
Yield of hybrid seed (kg/ha)	2000
Seed price paid to the seed growers (US$/kg)	0.60
Return to hybrid seed growers (US$/ha)	1200
Total cash costs (US$/ha)	800
Net return to hybrid seed growers (US$/ha)	400
Cost of seed production **with** subsidy (US$/kg)	0.40
Cost of seed production **without** subsidy (US$/kg)	0.60

Note: In addition to cash cost (paid-out cost), the Vietnamese government provided a subsidy to the seed growers of US$400/ha.

This is 30-35% higher than the profits from ordinary rice. However, the government's subsidy for hybrid rice seed production may not continue forever. If the subsidy is withdrawn, the procurement price of hybrid seed ought to be raised from its present level of US$0.6/kg to 0.8/kg, to compensate. This may lead to an increase of 30% in the selling price, over the current selling price of $1.1-1.2/kg. In order to keep the selling price at the current level without subsidy, the seed yield would have to increase to 3 mt/ha.

The production of hybrid rice seed must be economically viable if it is to attract seed companies. Therefore, seed production can be organised in countries where labor is available at a reasonable price, provided there is a demand for hybrid seed among farmers.

Future Demand

Successful development and large-scale dissemination of hybrid rice technology will have a major impact on the seed industry, in view of the fact that rice is a staple food crop in Asian countries. Usually, the private sector does not play much of a role in the early stage of technology development. At this stage, the public sector has to play a leading role.

Once the private sector is convinced that there is a demand for hybrid seed, it is willing to invest in commercial seed production, since hybrid seed production is quite profitable. Public sector research institutions in Asia

(IRRI and NARS) have already made great progress in the development of suitable rice hybrids, and have also developed many parental lines that are freely available to the private sector. A number of private seed companies (including 15 in India, six in Bangladesh, three in the Philippines, and one in Vietnam), and a few NGOs and farmers' organisations, have already begun to produce hybrid rice seed.

There is a growing linkage between the public and private sectors, not only in sharing genetic resources, but also in the form of private sector support for public research. Some small seed companies have also started joint ventures with big multinational companies for the production and/or import and marketing of hybrid rice seed in Bangladesh, Vietnam, the Philippines etc..

Some farmers' organisations have also begun hybrid rice seed production, either independently or as a joint venture with a private seed company. These developments may reduce the cost of hybrid rice seeds to as little as US$20-35/ha. This would promote the further adoption of hybrid rice production by small-scale farmers.

IMPACT ON RURAL EMPLOYMENT OPPORTUNITIES

In addition to its profitability, the additional labor cost of hybrid rice seed production is a positive feature in rural Asia. Hybrid rice seed production requires about 50 mandays/ha more labor than normal nonmechanised rice cultivation. It therefore generates additional employment opportunities. This is a great benefit, especially for landless women in countries such as India, Bangladesh, and Vietnam which have a labor surplus.

If hybrid rice technology spreads on a large scale in these countries, it would contribute substantially not only to food crop in Asian countries. Usually, the private sector does not play much of a role in the early stage of technology development. At this stage, the public sector has to play a leading role.

Once the private sector is convinced that there is a demand for hybrid seed, it is willing to invest in commercial seed production, since hybrid seed production is quite profitable. Public sector research institutions in Asia (IRRI and NARS) have already made great progress in the development of suitable rice hybrids, and have also developed many parental lines that are freely available to the private sector. A number of private seed companies (including 15 in India, six in Bangladesh, three in the Philippines, and one in Vietnam), and a few NGOs and farmers' organisations, have already begun to produce hybrid rice seed.

There is a growing linkage between the public and private sectors, not only in sharing genetic resources, but also in the form of private sector support for public research. Some small seed companies have also started joint ventures with big multinational companies for the production and/or import and marketing of hybrid rice seed in Bangladesh, Vietnam, the Philippines etc..

Some farmers' organisations have also begun hybrid rice seed production, either independently or as a joint venture with a private seed company. These developments may reduce the cost of hybrid rice seeds to as little as US$20-35/ha. This would promote the further adoption of hybrid rice production by small-scale farmers.

IMPACT ON RURAL EMPLOYMENT OPPORTUNITIES

In addition to its profitability, the additional labor cost of hybrid rice seed production is a positive feature in rural Asia. Hybrid rice seed production requires about 50 mandays/ha more labor than normal nonmechanised rice cultivation. It therefore generates additional employment opportunities. This is a great benefit, especially for landless women in countries such as India, Bangladesh, and Vietnam which have a labor surplus.

If hybrid rice technology spreads on a large scale in these countries, it would contribute substantially not only to food security, but also to poverty elimination by generating employment in rural areas. If six million hectares were to be planted in hybrid rice by the year 2010, hybrid rice seed production would generate about 3-4 million mandays of rural employment in Asia.

CONSTRAINTS TO HYBRID SEED PRODUCTION IN ASIA

Hybrid rice seed production is quite complex. Many factors affect the yield and quality of hybrid seed. When China began its commercial production of hybrid seed in the late 1970s, the average yield was less than 0.5 mt/ha. It took many years of experience to achieve its current seed yield of 2.5 mt/ha of F1 seed production in an area of about 0.14 million ha.

Twenty years after China, several other countries began the commercial hybrid production of hybrid rice seed. In a few small areas, seed yields of 4 mt/ha have been achieved (Ish Kumar, pers. comm.), but average yields still remain low at around 1.2 mt/ha. There are many constraints to the adoption of hybrid rice seed production technology. These are discussed below:

1. Hybrid rice technology in countries outside China was adopted only a few years ago. Seed growers need time to become familiar with the technology, and need more experience of hybrid seed production at specific locations under local climatic conditions.
2. There is a shortage of skilled manpower at the initial stages of seed production technology. Seed production can only be successful if it is

supervised by welltrained technicians and carried out by experienced growers.

3. Hybrid rice seed production is highly dependent on use of gibberellic acid (GA_3). Outside China this is quite expensive (more than US$1.00 per gram), because it is imported. In China it is quite cheap (US$0.30 per gram) because it is produced locally, so it does not cost much to apply a high dosage of this growth regulator.
4. The purification of parental lines is vital to the success of any hybrid seed production program. However, it has not received enough attention. Consequently, parental lines used for hybrid seed production are not pure, which results in poor-quality hybrid seeds.
5. Hybrid rice seed production requires a well-organised seed industry. Apart from China, most countries in Asia have not yet achieved this for hybrid rice seed.
6. Since hybrid rice technology is new in most countries, government policies do not yet motivate and encourage hybrid rice seed production on a large scale.
7. The linkage between public sector research institutes and seed production agencies working on hybrid rice is weak.

FUTURE OUTLOOK

Hybrid rice technology has had a tremendous economic impact in China, where it makes possible the production of about 20 million mt of additional rice each year. This extra production keeps global rice prices in check, so that the urban poor in riceconsuming countries can afford to buy enough rice. Without hybrid rice technology, the world would require 6 million ha more land to produce the same quantity. Thus, the technology has contributed not only to food security, but it has also helped indirectly to protect the environment.

The spread of hybrid rice technology would have the same effect in other Asian countries. Since hybrid seed production is labor-intensive, it would also create additional rural employment opportunities in countries with abundant labor.

Hybrid rice seed yields as high as 7.39 mt/ha have been reported from China, and up to 4.8 mt/ha from other countries. Seed yields can be increased by bringing about:

1. Further genetic improvement of flowering behavior and floral traits of seed (e.g. exserted stigma) and pollen parents (e.g. abundant pollen);
2. Modification of seed production practices; *and,*
3. Selection of the most favorable seasons and locations for seed production.

GA_3 application is an important component of hybrid rice seed production technology. Its cost is very high in many countries. High seed yields in China have been achieved with a very high dosage of 150-300 grams per hectare. Outside China, it is essential to reduce the cost of GA_3, either by reducing the dosage or by producing cheap GA_3 locally. Alternatively, a cheaper substitute for GA_3 needs to be found.

The discoloration of hybrid rice seeds caused by tropical fungi is an important problem which needs to be tackled. In China and northwestern India, CMS lines in hybrid rice seed production plots have been found to have a higher incidence of seed-borne diseases (such as paddy bunt, caused by *Neovassia horinda,* and Tak and False Smut caused by *Ustilogonoids virens*) compared to the pollen parents. This can cause a serious outbreak of these diseases in commercial crops of hybrid rice, and therefore needs more attention.

For countries such as Japan, Korea, Malaysia and United States which have a labor scarcity and high wages, hybrid rice seed production needs to be mechanised to make it cost effective. A mechanised hybrid rice seed production system has been developed in the United States. Such a system in Asia might use several strategies, such as:

1. The use of facultative female sterility.
2. Incorporating a gene for herbicide susceptibility in the pollen parent.
3. The use of highly heritable differences in grain width (at least 0.7 mm) in seed and pollen parents for mixed planting and mechanised harvesting, and.
4. Incorporation of a phenol reaction (Ph) gene into the pollen parent.

For countries lacking suitable seed industry infrastructure, a self-sustaining seed production system can be tried, so that farmers can produce hybrid seeds for their own use rather than for sale to others.

Apomixis is the ultimate genetic tool to develop true breeding hybrids and facilitate commercial exploitation of heterosis by even resource poor farmers. (Apomixus is reproduction involving specialised tissue but not dependent apon fertilisation: an example is parthenogenis). So far, there is no confirmed report of apomixis in rice. However, research is in progress in Mainland China, at IRRI, and elsewhere, to discover, induce, and/or genetically engineer such rices.

REFERENCES

He, G.I., X. Zhu and J.C. Flinn. 2019. Hybrid seed production in Jiangsu province, China. *Oryza* 24: 285-296.

Janaiah, A. and M. Hossain. 2020. Hybrid rice for food security in the tropics: An evaluation of farm-level experiences in India. Paper presented at the 3rd

International Crop Science Congress held in Hamburg, Germany, 17-22 August 2020.

Mao, C.S. 2018. Hybrid rice seed production in China. In: *Seed Health*. International Rice Research Institute, Los Banos, Laguna, Philippines. Pp. 277-282.

Mao, C.X. 2021. Improving seed production to speed up the global commercialisation by hybrid rice. 2001. In: *Rice Research* for Food Security and Poverty Alleviation, S. Peng and B. Hardy (Eds.). Proceedings of the International Rice Research Conference, 31 March-3 April 2020. International Rice Research Institute, Los Banos, Laguna, Philippines. 692 pp.

Maruyama, K. and K. Oono. 2021. Induction of mutation in tissue culture and its use for plant breeding. VII. On the female and male sterile rice plants regenerated from seed acalluses. *Jan. Jour. Breed.* 33 (Suppl. 1): 24-25. (In Japanese).

Xu, S.J. and B. Li. 2019. Managing hybrid rice seed production. In: *Hybrid Rice*. International Rice Research Institute, Los Banos, Laguna, Philippines, pp. 157-163.

Yuan, L.P. 2015. A Concise Course in *Hybrid Rice*. Hunan (China): Hunan Science and Technology Press. 168 pp.

4

Seed Production of Maize and Millets

MAIZE SEEF PRODUCTION

Maize is a cereal grain domesticated in Mesoamerica and subsequently spread throughout the American continents. After European contact with the Americas in the late 15th and early 16th centuries, maize spread to the rest of the world.

Maize is the most widely grown crop in the Americas (332 million metric tons annually in the United States alone). Hybrid maize, due to its high grain yield as a result of heterosis ("hybrid vigor"), is preferred by farmers over conventional varieties. While some maize varieties grow up to 7 metres (23 ft) tall, most commercially grown maize has been bred for a standardized height of 2.5 metres (8 ft). Sweet corn is usually shorter than field-corn varieties.

The maize plant has separate male and female flowers. The male flowers, or tassels, are located at the top of the plant, whereas the female flowers (ears and silks) develop about half way up the stalk. The location of the tassel at the top of a relatively tall plant and its separation from the female flower promotes cross-pollination between plants. Commonly, pollen from a tassel will be blown by the wind to one or more plants nearby. Pollen is very small, and a single tassel may produce up to 25 million pollen grains. Shortly after pollen contacts the female silks, it germinates and grows down through the silk, eventually fertilising the young ovule. The fertilised ovule develops into the embryonic plant within the developing seed.

Maize stems superficially resemble bamboo canes and the internodes can reach 20–30 centimetres (8–12 in). Maize has a very distinct growth form; the lower leaves being like broad flags, 50–100 centimetres long and 5–10 centimetres wide (2–4 ft by 2–4 in); the stems are erect, conventionally 2–3 metres (7–10 ft) in height, with many nodes, casting off flag-leaves at every node. Under these leaves and close to the stem grow the ears. They grow about 3 milimetres a day.

The ears are female inflorescences, tightly covered over by several layers of leaves, and so closed-in by them to the stem that they do not show themselves easily until the emergence of the pale yellow silks from the leaf whorl at the end of the ear. The silks are elongated stigmas that look like tufts of hair, at first green, and later red or yellow. Plantings for silage are even denser, and achieve an even lower percentage of ears and more plant matter. Certain varieties of maize have been bred to produce many additional developed ears, and these are the source of the "baby corn" that is used as a vegetable in Asian cuisine.

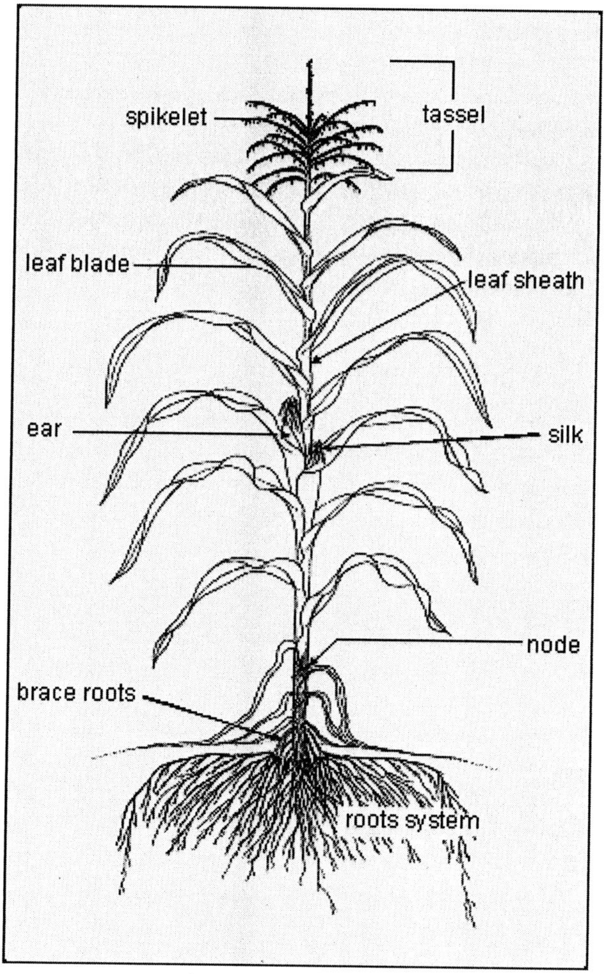

Figure 1. Maize plant

Maize is a facultative long-night plant and flowers in a certain number of growing degree days > 50 °F (10 °C) in the environment to which it is adapted. The magnitude of the influence that long nights have on the number of days that must pass before maize flowers is genetically prescribed and regulated by the phytochrome system. Photoperiodicity can be eccentric in tropical cultivars, while the long days characteristic of higher latitudes allow the plants to grow so tall that they do not have enough time to produce seed before being killed by frost. These attributes, however, may prove useful in using tropical maize for biofuels.

The apex of the stem ends in the tassel, an inflorescence of male flowers. When the tassel is mature and conditions are suitably warm and dry, anthers on the tassel dehisce and release pollen. Maize pollen is anemophilous (dispersed by wind) and because of its large settling velocity most pollen falls within a few meters of the tassel. Each silk may become pollinated to produce one kernel of maize. Young ears can be consumed raw, with the cob and silk, but as the plant matures (usually during the summer months) the cob becomes tougher and the silk dries to inedibility. By the end of the growing season, the kernels dry out and become difficult to chew without cooking them tender first in boiling water. Modern farming techniques in developed countries usually rely on dense planting, which produces one large ear per stalk.

Seeds

The kernel of maize has a pericarp of the fruit fused with the seed coat, typical of the grasses. It is close to a multiple fruit in structure, except that the individual fruits (the kernels) never fuse into a single mass.

Figure 2. Maize seeds

The grains are about the size of peas, and adhere in regular rows round a white pithy substance, which forms the ear. An ear contains from 200 to 400 kernels, and is from 10–25 centimetres (4–10 inches) in length. They are of various colors: blackish, bluish-gray, purple, green, red, white and yellow. When ground into flour, maize yields more flour, with much less bran, than wheat does. However, it lacks the protein gluten of wheat and, therefore, makes baked goods with poor rising capability and coherence.

Choice of Maize OPVs

Maize breeding programs are continuously developing new OPVs. They are often higher yielding than older varieties or farmers' own varieties and may have other value-added traits, such as early maturity or better disease resistance.

To benefit from a new, improved variety, farmers must first obtain seed of the variety. Seed production must then be done in a particular manner that maintains the purity of the variety and gives good quality seed. This is normally done by requesting foundation seed from either the national maize program or an international center such as CIMMYT. Alternatively, foundation seed may be available through private sector sources.

When producing seed, a farmer usually wants to maintain the characteristics of a variety. Crosspollination between different maize varieties must therefore be prevented. Isolating the seed production field from other maize fields helps achieve this. If two different varieties are grown next to each other, cross-pollination will occur between the two varieties, and the crop grown from such seed will have a mixture of the characteristics of those two varieties.

Isolating a Seed Production Field

Isolation of a seed crop can be done in four ways:

1. By space. Maintain a distance of at least 300 meters between the seed crop and any other maize field that has a different variety planted.
2. By time. Sow your seed crop a month earlier or later than neighboring maize fields.
3. By certified seed barriers. A barrier of genetically pure seed of the same variety may be planted within the isolation distance of the seed production field.
4. By natural barriers. Seed production plots can be established on land isolated by natural or artificial forests, for example.

Distance and time isolation are the most commonly used approaches. The goal is to have no other maize variety shedding pollen nearby when the seed production field is flowering. Wind may carry pollen further than 300 meters. Thus, if there are constant strong winds in one particular direction, the distance to the next maize field should be at least 400 meters.

How to Select a Suitable Field

Consult with your neighbors about when and where they will plant their maize, so you can sow yours in a field that is properly isolated. Apart from considering isolation, select your best field for maize seed production and manage it well, because the value of good seed is higher than the value of grain. Choose a field where no maize has been grown during the previous year to reduce the possibility that last year's maize crop may germinate and cross-pollinate your seed maize. Farming communities may want to produce the seed of one variety for the entire community in a single field. This will help in finding a wellisolated field with good soil. Also, the village can make a collective effort to look after the maize crop well.

How to Manage your Seed Production Field

Prepare the soil in your field at least two weeks prior to planting. If there are any grains from previous maize crops left in the soil, they may germinate in these two weeks and you should remove them when you sow your maize seed crop.

If possible, sow your maize seed crop early. Choose a plant density that is recommended for your area. A common seeding rate for maize is 20 kg/ ha. Planting in straight rows will facilitate crop management and seed production operations. Apply fertiliser and remove weeds in time.

Eliminating Undesirable Plants

Carefully examine your maize seed crop as it grows. You may find plants that look very different or flower much earlier or later than most other plants in the field. These plants are called off-types and they should be removed—a process known as "roguing"—before pollen shedding starts. Most farmers do not like to remove maize plants from their field, but rouging is critical to maintain varietal purity.

Harvest and Drying

During harvest and drying, be careful that the seed of your maize crop does not get mixed with seed or grain from other maize varieties. Keep only the best and healthiest ears and kernels for seed, and use the rest of the harvest as grain. Your best seed comes from healthy undamaged ears that are typical for the variety. Thus, discard off-types, rotten and damaged ears, and ears where the kernels have started to germinate or are affected by insects. Place the harvested ears on a clean and dry surface, such as concrete or plastic, and dry them well in the sun. To make sure that all kernels get exposed to the sun, spread the ears in a flat layer and turn them several times. Maize seed stores best at <12% moisture content.

Shelling, Cleaning, and Storing Seed

When the maize seed is dry, it can be stored on the cob or shelled. Be careful not to damage the seed when shelling. Your best seed typically comes from the middle part of the ear. After shelling, clean the seed, removing dirt and other inert matter. Remove seeds that are small, look diseased, have started to germinate, or are damaged by insects. Treat the seed with an insecticide-fungicide combination, and store it in a dry and cool place. When treating the seed, follow the guidelines on the seed treatment package carefully. One good way to store seed is to place it in a jute bag, close the bag and drop it into a plastic bag, close the plastic bag and place this in another jute bag, finally closing the outer jute bag. Seed bags should be stored in a seed warehouse on wooden pallets in cool conditions away from fertiliser and chemicals.

Quality Control in Maize Seed Production

The production and distribution of quality maize seed requires diligent efforts both during field production and post-harvest handling. Field inspections are commonly conducted at different crop development stages to ensure quality.

Field Inspections

A planting inspection is commonly conducted to determine that the maize seed planted is genetically pure, of known origin, and is an appropriate variety for the area. If a mechanical planter is being used, it should be checked to ensure that it is clean and free of maize seed of other types and is properly calibrated to achieve recommended seeding densities. The field should be inspected to verify that it is properly isolated and free of volunteer plants.

A second field inspection may be made during the vegetative growth phase. Isolation should be checked, along with the presence of disease, insect pests, or weed infestations. At this stage, off-type and diseased plants may be rogued.

The most important field inspections are made just prior to and during flowering. At this time the maize seed field is most susceptible to genetic contamination from wind-blown pollen coming from off-type plants within the field or other maize varieties in surrounding fields. Therefore, it is essential during the pre-flowering inspection to confirm that the maize seed field has been properly rogued and is sufficiently isolated from other maize fields. Plants that are off-type or diseased, along with harmful weeds, must be removed at this time.

A pre-harvest or harvest inspection may be conducted as the crop reaches maturity and the seed has lost a significant portion of its moisture content. Off-type plants, such as those that are still green when most the other plants are dry, may be removed at this stage. At harvest, ears with different grain color or texture from the produced variety should be removed.

Quality Control Tests

Various standard tests for moisture content, germination, and physical purity can be conducted to evaluate the quality of the seed. The most common evaluation described here is the germination test, designed to determine the seed's capacity to germinate and produce normal plants when sown under appropriate conditions.

Germination Test

Germination tests may be conducted by building an open wooden box, 1 m long, 50 cm wide, and 10 cm deep. Fill the box with insect-free, loose soil. Divide it in half and plant 100 seeds in rows separated by 10 cm. Since the objective is to see how many seeds germinate, sow them one by one in a thin line about 2 cm deep. The box should be watered thoroughly and kept in a safe place away from birds and other animals. Another alternative would be to conduct the germination test in a well-prepared bed near the homestead or in a protected garden. Another option is to evaluate maize seed germination in paper towels or newspapers. With this method, 50 maize seeds are spread out in lines on moistened paper and then covered by another wetted paper. The paper is rolled and tied with a string or elastic band. The paper rolls can be placed in plastic bags or other containers and kept moist for seven days. With each of these methods, a count for normal developing seedlings is made between 7 and 10 days after sowing. A minimum of 80% germination is the suggested standard; anything below this means the seed is not of acceptable quality.

SEED PRODUCTION OF SORGHUM AND PEARL MILLET

Sorghum and pearl millet are important indigenous African cereal food crops. To date, several improved varieties and a number of agronomic recommendations have been developed for different categories of farmers in different agro-ecological regions. These include open-pollinated varieties (OPVs) and hybrids for grain and forages; and recommendations on spacing, plant population, and fertiliser rates. The improved varieties are diverse in maturity, adaptation, height, seed color, and size. Generally, they are tolerant to prevailing diseases such as anthracnose, downy mildew, ergot, smut, viruses, leaf diseases, and sooty stripe.

Due to numerous constraints, seed production and supply for traditional food crops such as sorghum and millet are unstable and often lack continuity. As a result, promotion and adoption of new varieties in

Zambia is limited. Statistics on the production and sale of sorghum and pearl millet seed/grain are not readily available. Singh and Jain estimated national seed requirements at 336-480 tons for sorghum and 295 tons for pearl millet. During 1985-94, area planted to sorghum and millet grew by 0.2 and 13.1% per year, respectively. Thus, current seed requirements could be higher than this estimate. Also, the estimate was made when only a few improved varieties were available on the market. As new varieties spread and adoption increases, seed demand generally increases. Farmers are increasingly aware of the benefits of improved seed and of differences among cultivars. In particular, farmers value the early maturity, large seed size, and higher grain yields of new varieties. Guidelines for producing quality seed of these varieties have also been developed. The recommendations are similar to best agronomic practices for the respective varieties, although certain additional procedures are suggested, depending on the class of seed.

Figure 3. Sorghum seeds

Figure 4. Pearl millet seeds

Seed Production Procedures

Procedures governing production and sale of seed need to conform to the national seed act. Standards have to be met for different crops and seed classes, and are enforced by licensed personnel affiliated to the Seed Control and Certification Institute (SCCI), which controls the Seed Certification Scheme. Table 1 shows some of the requirements for seed production of sorghum and millet.

Table 1: Seed classes and production restrictions for sorghum and pearl millet.

Sorghum*		Pearl millet**	
Seed Class	Isolation distance, m	Seed Class	Isolation distance, m
Basic OPVs	200	Nucleus	2,000
Certified OPVs	100	Breeder	1,000
Basic hybrid	300	Basic	1,600
Certified hybrid	200	Certified	300 – 400
Basic forage	400	Inbred lines	800
Certified forage	400		
Inbred lines	800		

* Adapted from Nath 1995. ** Adapted from Singh 1995.

Nucleus and Breeder Seed

Usually handled by the breeder, these are the first stages in the chain of seed production. The breeder has strict control, observes isolation distances, inspects the crop, and rogues off-types regularly to maintain genetic purity. Sufficient quantities of seed should be made available by the time a new cultivar is at the pre-release stage. OPV breeder seed is produced under bulk planting, but usually in small plots with wide spacing to allow for maximum expression and roguing.

Seed production for a hybrid is highly technical and requires experience. Four isolations are normally maintained as follows: A-line x B-line to increase seed of the male-sterile A-line. Separate isolations are required to increase the B-line and R-line. Finally, hybrid seed is produced by crossing an A-line x R-line. Synchronous flowering of hybrid parents (nicking) is mandatory, as is rigorous rogueing of off-types and pollen shedders in the A-line.

Hand pollinations and selfing in the B-and R-lines is possible. Different row ratios of hybrid parents of 4:2 or 6:2 (A-line against the B- or R-line) are used with 4-8 border rows of B/R lines planted all round the field. Stagger planting of the B-line is possible but could present problems for the R-line. This category of seed is generated every 3-6 years, either during the normal season or in the off-season, under irrigation.

Basic (Foundation) Seed

Part of the Nucleus or Breeder seed is used to produce the Basic or Foundation Seed, either by researchers or a few contracted seed growers. Similar guidelines are followed as those for the first class of seed. The seed crop must be inspected at least three times by inspectors affiliated with the SCCI: during the vegetative growth stage, at flowering, and at maturity.

Certified Seed

Basic seed is used to produce certified seed under a similar planting pattern to Breeder and Foundation seed. For hybrid parents, however, rows of the B-line are replaced by the R-line.

Training Seed Producers

Some training is given to seed growers. Training courses are offered to individuals representing groups, organisations institutions. Sometimes field days are conducted to improve awareness of the improved cultivars available and illustrate important aspects of seed production (field selection, planting pattern, weeding, and fertiliser rates, among others).

Site and Field Selection

Locations and fields for seed production may differ from those for commercial crop production. Important consideration should be given to growing conditions, length of rainy season, irrigation facilities, humidity during ripening period, temperatures and daylength, protection from strong wind, adequate isolation distance, and the presence of birds, diseases, and insect pests or related wild/cultivated species. The field should be relatively fertile, free from water logging, and with a pH not.

Field Preparation

Sorghum and pearl millet seeds are small. Soil should have fine tilth, and be moist and weed free at the time of planting. Zero tillage is not recommended.

SOWING

Drilling or stations should be done in rows at a depth of 2-3 cm when rains are established, and in such a way that the seed crop matures during a period with minimum precipitation. It is important to know the maturity period of the cultivar.

Off-season planting under irrigation is possible, especially in hot valley areas. For hybrid parents, planting should be in such a way that they flower at the same time. Recommended seed rates are 4-6 kg/ha for pearl millet and 8-10 kg/ha for sorghum to achieve plant populations of 60,000-80,000 and 130,000-150,000 respectively. The recommended spacing for pearl millet is 75 cm between rows and 20 cm between hills within a row or 60 x 60 cm. For sorghum, 60-75 cm x 50 cm is recommended, with 4-8 seeds per station, later to be thinned to 2-3 plants per station. It is advisable for a block of farmers to plant the same variety at the same time. Dry planting is not recommended.

Fertiliser Rates

200 kg/ha of Compound D as a basal fertiliser and 100 kg/ha of urea as top dressing are recommended. Top dressing can be applied in two splits. Fertiliser applications will differ according to local soil type and fertility.

Weeding

Seed plots are normally weeded 2-3 times using hoes, oxen, or tractor-drawn cultivators to keep fields weed-free at all times. In sorghum a combination of mechanical weeding and pre-emergence Gesaprim herbicide should be applied at 3-4 l/ha. Weeds compete with the seed crop for nutrients, sunlight, and soil moisture. They could also be a reservoir of diseases that contaminate the crop at harvest.

Thinning

Thinning to the recommended spacing and plant population is required before tillers form. Gap filling is not recommended in a commercial seed crop.

Plant Protection

Improved sorghum and pearl millet cultivars are tolerant of most prevailing diseases. Avoid growing seed in disease-endemic areas. Bird damage, potentially the most serious problem, can be minimised by careful selection of the site, growing larger areas, scaring, and planting varieties with some "resistance" to birds (for example, brown high-tannin sorghum and bristled types of pearl millet).

Roguing and Inspection

Roguing is done to remove weak, diseased and off-type plants before they shed pollen, during flowering, and before harvest.

Harvesting, Threshing and Post-Harvest Handling

Hand harvesting is the most common practice. Combine harvesting is possible for short stature varieties and is done after final inspection. Harvesting should be done when the seed crop is fully mature but before lodging. Mechanical mixtures should be avoided when harvesting hybrids and hybrid parents, especially when lodged plants are present. Male rows should not be used for seed. Trained workers can remove left-over undesirable plants during harvesting. Manually harvested panicles are sun dried for a few days and then threshed using suitable mechanical threshers, mortar and pestle, or sticks (beating). Care must be taken to avoid breaking or cracking the seed. Clean the threshing equipment before using it on a different variety or seed class. Clean the threshed seed by winnowing, then treat with available chemicals and store in screw-cup containers, bags or tins. It is important to record the identity of each seed type and class. Seed classes are given labels of specific colors.

Seed Marketing and Pricing

Seed marketing and pricing depend on various factors, including type of seed, demand, the producers themselves, and sometimes distance from production and selling points. Prices may vary across different crop species, between hybrid and OPV seed, and across seasons.

REFERENCES

AfricCOMMS (PVT) Ltd. 2022. *Open-pollinated maize seed production handbook.* Harare, Zimbabwe: SADC/GTZ.

Beck, D.L. 2024. "Hybrid corn seed production". In C.W. Smith, J. Betran, and E.C.A. Runge (eds.), *Corn: Origin, History, Technology, and Production.* Hoboken, New Jersey: John Wiley & Sons, Inc.

Chopra, K.R., R. Chopra, G. Rabbani, and K.K. Thimaiah. 2019. "Seed production". Pp. 445-477 in *Pearl Millet Breeding*. New Delhi, India: Oxford and IBH Publishing Company Pvt. Ltd.

FAO. 2016. *The world sorghum and millet economics-facts, trends and outlook.* Patancheru 502 324, Andra Pradesh, India: International Crops Research Institute for the Semi-Arid Tropics; and Rome, Italy: FAO.

Heemskerk, W. 2021. *Seed selection and seed storage.* Ministry of Agriculture.

Nath, B. 2015. Sorghum (*Sorghum bicolour* L). Pages 165-171 in Zambia seed technology handbook. Ministry of Agriculture, Food and Fisheries.

Singh, P. 2015. Pearl millet (*Pennisetum americanum* L.) pages 171-182 in Zambia seed technology handbook. Ministry of Agriculture, Food and Fisheries.

The Maize Program. 2019. Development, Maintenance, and Seed Multiplication of Open-Pollinated Maize Varieties, 2nd edition. Mexico, D.F.: CIMMYT.

5

Bean Seed Production

The common bean (*Phaseolus vulgaris*) is a member of the *Fabaceae* or legume family and is of New World origin. The wild ancestors of the modern common bean come from Central and South America. These ancestral types are found across a range of environments, from moderately hot, arid climates to humid lowland tropics and even into cooler upland areas of South America. The beans of this species grown in North America today are grown in a more limited temperate climatic zone. Modern agricultural varieties of snap beans and kidney beans as well as some navy beans and cranberry beans have a non-climbing (bush-type) habit.

Most other agricultural varieties have a vining habit classified as either a type II (upright short vine) or a type III (floppy vine). These types are indeterminate in their growth habit, but are not climbing. Most navy and black varieties are type II, while most pinto, red, and great northern varieties are type III. Older varieties, landraces, and heirlooms are often indeterminate and climbing, and are referred to as "pole beans". The snap bean types, also called string beans, green beans, and garden beans, are grown for a harvest of tender young pods with immature seeds eaten as a fresh vegetable. While immature pods of any common bean variety were once commonly eaten, there are now thousands of varieties developed for fleshy pods with slow seed maturation for a crop that has become a vegetable mainstay in our culture.

The term "snap bean" refers to the way in which fresh garden beans are broken or snapped by hand into short segments before cooking. In the bean seed growing regions of the Northwest these are called "garden beans" to distinguish from the more widespread dry beans or field beans. Whether for vegetable or seed, beans are produced as an annual crop maturing in one growing season. Bean seed crops may require a long season to mature (average 90-120 days) and must be planted early enough to mature and dry prior to fall frosts or rains. Commercial bean seed production is focused in the western region of the United States as the dry, long summers and arid climate create good growing conditions with lower disease pressure and a long season for maturation.

Figure 1. Common bean plant

GROWING BEAN SEED

Climatic Requirements

Common bean is a tender, warm season crop that requires warm, well drained soils for germination. Temperatures of 70°–80°F (21°–27°C) are preferred for optimum crop growth. Temperatures above 90°F or below 50°F during flowering may adversely affect pod set and seed yields. Most snap bean cultivars germinate best when soil temperatures are at or above 65°F (12°C.), but germination may be inhibited at temperatures above 95°F (35°C). There are instances when seed growers must plant with soil temperatures below optimum in order to fully mature a seed crop by the end of the season. Cultivars vary considerably in their ability to germinate in cool, moist soils and to resist common root rot organisms that can damage or destroy seedlings. Varieties with green cotyledons in the dry seed and white cotyledons at emergence have what is called the "persistent chlorophyll" or pc trait. Varieties with the pc trait, including most French filet types, green-seeded snap beans, and a number of modern snap bean varieties, are the most sensitive to cool soil conditions. Second in sensitivity are white-seeded dry and snap bean varieties, while colored seeded varieties are the least sensitive to cold and moisture stress at emergence. At Oregon State University, Oregon bush blue lake varieties have been selected for good germination and emergence in cool, wet soils.

Soil and Fertility Requirements

Soil conditions and types may vary and still produce healthy plants and superior seed, but adequate drainage is essential as beans are sensitive to both moisture stress and water logging. Early plantings for long season varieties are best done on light sandy soils. Soils that easily crust should be avoided and irrigation and tillage managed to avoid crusting as seedling emergence may be impaired. All crop residues should be thoroughly incorporated into the soil prior to planting. Residues on the surface can interfere with seed placement during drilling and result in inadequate coverage of weeds with soil during an early cultivation. Incorporation of residue is additionally crucial in the availability of nitrogen (and other nutrients) to the young bean plants before their development of symbiosis with the *Rhizobium* bacteria. Some growers inoculate their bean seed with *Rhizobium* innoculant prior to planting. This practice can be particularly beneficial when producing on soils where beans have not been grown in the past or that are low in biological activity. It is important, however, for organic producers to use non-genetically modified sources of innoculant.

Planting and Cultivation

Common beans are frost sensitive. The usual recommendation is to plant after all danger of frost is past. However, it is often necessary to plant a seed crop 7-10 days before the last anticipated frost either to allow time for full maturity and harvest prior to inclement weather in the fall or to avoid the heat of summer during flowering (see *Flowering and Pollination*). Soil temperatures do need to be warm, approximately 65°F (18°C), so the seed germinates quickly and does not succumb to root rots in cool soils. However young plants with fleshy cotyledons are more frost tolerant than mature bean plants and can tolerate a light frost soon after emerging.

Bean seed is commonly planted at an average depth of 1–1.5 inches (2.5–3.8 cm) below the surface of the soil after the press wheel has packed the soil. However, planting depths may vary according to soil moisture and local conditions. In the Willamette Valley of Oregon bean seed is often planted at ½ to ¾ inch depths while in southern Idaho 2-3 inch depths may be required to attain adequate soil moisture. Common beans for seed are routinely grown on 22 inch row centers with an in-row density of 6 to 8 bean plants per foot. Seed may be inoculated with *Rhizobium* bacteria prior to planting to enhance nitrogen availability. This may be particularly beneficial when planting on soils where beans have not been produced in recent years. Prior to planting beans soils are commonly pre-irrigated, allowed to dry to moderate soil moisture, planted, and then not irrigated again until seedling emergence. This avoids soil surface crusting prior to seedling emergence.

Alternatively soil crusting is mitigated by a very shallow dragging of soil with a harrow prior to emergence. Post-planting irrigation is not recommended because the sudden shock of cold water causes mechanical damage to seed, leading to reduced germination and emergence.

Weed control is particularly critical in organic bean seed production to achieve optimum growth and prevent weed seed contamination in harvested seed. Control of hairy night shade (*Solanum sarrachoides*) is particularly critical as the berries of this weed are very difficult to remove from harvested seed. Weed management and cultivation practices may vary according to farm equipment availability, planting practices, and local field conditions. However, one successful method that has been used for weed control under organic production is as follows: Plant single rows into hills that are 6-8 inches (15-20 cm) high. After 4 to 5 days, when seeds have sprouted but the epicotyl is still 1 inch (2.5 cm) below ground, cultivate the hills for the first time using a drag harrow. The drag harrow will scrap off the top surface of each hill, exposing the first set of tiny weeds before they can become established in the bean rows. Even if done properly an occasional bean plant is lost, but it is well worth the effort to get the first "within row" weeds.

The next cultivation is done after the bean plant has emerged and has one set of true leaves. Set up two tool bars on a cultivating tractor. The first tool bar should have shanks with chisel points that are set as close to the bean rows as possible. These chisel points will disturb the soil, killing weeds close to the plant, as well as throwing some soil up onto the base of the plant to smother other weed seedlings.

Follow these shanks by a second tool bar with "duck feet" sweeps which will disturb the soil, kill weeds between the rows, and throw soil back onto the hills to rebuild them. Subsequent cultivations are done with knives that cut soil away from the bean plants, and sweeps between the rows to rebuild the hills. In addition to disturbing as much of the soil surface as possible to up-root or bury weed seedlings, it is important to create sufficient hills in order to facilitate undercutting the bean plants at harvest. Cultivation should be done with attention to avoiding major root disturbances as the plants get larger. Many growers do not use the described drag harrow method, but practice several cultivations and additionally walk the field to remove larger weeds once the canopy is closed.

Flowering and Pollination

Nearly all modern agricultural varieties of common bean grown at higher latitudes are dayneutral in their flowering response. In other words, the length of day does not determine flower initiation and plants flower as soon as they are physiologically ready. Many tropical and some temperate dry bean varieties are day-length sensitive. Many tropically adapted common beans are short day plants and fail to flower until the days shorten in late summer. As varieties vary in flowering response it is recommended to grow a small plot of any new seed crop to see if it matures adequately in the region.

Common bean flowers have the shape and structure of a typical legume flower. Their petals are usually whitish, but may be tinged to deep violet to purple. There are five petals consisting of one large, oval banner or standard petal, two elongate keel petals that are fused together enclosing the stamens, and two wings. There are nine stamen filaments united into a sheath that surrounds the pistil with one stamen filament separate from the fused nine. Unlike other legumes, the style and keel petals of a bean flower follow an approximately 360° spiral pattern. This trait is one that taxonomically separates *Phaseolus* from all other genera.

Common beans are highly self-pollinated as the anther sacs are borne directly adjacent to the stigma (the receptive part of the female flower) and the pollen is normally released the day before the flower opens. Under most conditions the seed is self-pollinated before the flower opens and is accessible to insect activity. However, there are several situations that can lead to out-crossing between bean plants including: 1) When growing beans adjacent to areas with biologically diverse habitats for insects there is usually a higher and more diverse number of insect pollinators present. These pollinators may be able to get into flowers before they fully open either because the flowers are not completely closed (a genetic trait) or the insects may be able to pry the flowers open in search of nectar or pollen; 2) Variety differences of floral structures exist and some (particularly pole bean varieties) have longer pistils which extend further out of the flower making cross pollination more likely; 3) An increase in rate of crossing can occur due to a combination of excessively hot weather and the fact that some bean varieties are more sensitive to heat than other varieties. The pollen of many bean varieties can be damaged and non-viable if temperatures are too high at the time of flowering while the stigma will usually remain unaffected. There is genetic variation for this condition among bean varieties and the pollen of one variety may loose viability while another does not under similar conditions. If multiple varieties are being produced during a hot period it is possible that pollinating insects may move pollen from a fertile variety to another variety lacking fertile pollen and create unwanted crosses.

ISOLATION REQUIREMENTS

Isolation requirements can vary depending on location and type of production. If conditions are conducive to an increased rate of cross pollination (see *Flowering and Pollination*) then isolation distances should be increased. The common recommendations for isolation are that little or no isolation is necessary, and under most situations even with heat and ample pollinators there is only a small percentage of outcrossing that occurs. In commercial production a break of 10–15m (30–45ft) is recommended with a taller crop barrier, such as several rows of corn or sunflowers, to deter pollinator movement. For foundation or stockseed where genetic purity is critical this break should be increased to 50m (150ft) with a tall barrier crop as well. Vining or "pole" varieties of common bean are more prone to outcrossing than bush varieties and the recommended isolation distance should be doubled.

GENETIC MAINTENANCE AND IMPROVEMENT

Population Size

Common bean is a self-pollinated species that has evolved to tolerate inbreeding. As such any commercial seed crop is ample in scale to avoid inbreeding depression. Population size becomes an issue however from a genetic conservation perspective when saving seed from a small population or conducting a selection within a larger population. In addition to the importance of genetic preservation diversity is important when breeding or selecting varieties for adaptation to varied environmental conditions. While maintaining diversity may be desirable in these instances, it must be remembered that routine selection, therefore narrowing the population's diversity, is critical in order to avoid unwanted off-types and maintain integrity of a commercial variety (see *Selection criteria*). Diversity will naturally arise in a population that is not routinely selected and may include undesirable traits such as fibrous and stringy pods resulting in an inferior commercial product. The following recommendations are intended as a guide to ensure preservation of desirable traits for conservation and breeding purposes.

Most modern bean varieties are derived from a single elite plant that exhibits all of the characteristics of the breeder's ideotype. Seed multiplied from this single plant is what plant breeders call a "pure line" and the subsequent variety is a "pure line variety." A variety derived in this way has very little inherent genetic diversity and therefore seed saved from less than 10 true-to-type plants is sufficient to ensure all of the variety's genetic diversity is conserved.

When reproducing an older bean variety, the question of genetic diversity needed to maintain the genetic breadth of a variety is more complex. The older commercial bean varieties from the 1920s through the 1970s have now been around long enough that they have probably widened their genetic variability since they were first released as pure line varieties. For this reason it is wise to save seed from at least 50 representative plants whenever reproducing an older commercial variety.

Lastly, there are also older farmer varieties, land race varieties, and heirlooms. These populations often have more genetic variation than commercial varieties. As this variation may be very important in the environment or the cultural context in which these varieties are used, it is important to maintain this potential genetic breadth by saving seed from at least 100 representative plants (200 if the variety is a highly variable land race).

Selection Criteria

Selection, or roguing, refers to removing unwanted off-types in the population. Routine selection is critical to maintain a variety's trueness of type and avoid the appearance of undesirable traits. In commercial seed production, selection should be toward the seed company's variety description or the breeder's original vision of the variety. Selection in all varieties should include quality traits such as tender pods in snap beans. Additionally, it is critical to select against disease in bean crops in order to produce quality seed, minimise risk of seedborne diseases, and select for or maintain a cultivar's disease resistance (see *Disease).*

There are very few stages in snap bean growth in which differences within a variety are apparent. If the seed is planted into uniform soil tilth and moisture then it is possible to perform early selection for speed of emergence, uniformity of the stand, and overall vigor, which are essential traits for organic production. By routinely roguing out late emerging, low vigor seedlings there will be improvement over cycles of selection for these traits. For some traits, like leaf shape, color, and plant stature it is possible to evaluate the plant and rogue it to type before flowering has begun. Vining plants in a non-climbing, bush variety should be selected against. Flower color, placement, timing and duration of flowering can all be selected during flowering. Pod characteristics such as color, shape, and length are also notable and should be selected.

For snap beans it is particularly important to select against plants with off type pods. Off-types may have altered pod shape (oval rather than round), increased pod fiber, or the presence of strings in a stringless variety. Pods need to be examined at the green pod stage. High fiber pods are often recognisable because they are smaller, oval in shape and curved. Stringy pods can be identified by pressing on the "spur" (the remnants of the style). "Hardpoint" spurs indicate strings while soft, flexible spurs are associated with stringlessness. Off-type and normal pods may be found on the same plant, so it is important to examine all of the pods on every plant.

HARVESTING

Proper timing of harvest is important in order to produce high quality bean seed that is fully mature, has a high germination percentage and has maximum storage potential. Each variety has its own specific harvest timing and while this makes overall recommendations for gauging cutting, curing, and threshing difficult, there are basic signs that indicate maturity. The initial signal that the crop is ready to cut is the relative maturity of the pods and their color at or near the time when they are breaking, or when they first turn yellow, mahogany, or "buckskin" in color but haven't gone to the tan, papery shade. Maturation to the brown, papery stage increases danger of seed shattering during harvest. Pods should generally be yellow at harvest in order to mature properly in the windrow, but the exact desired color may be variety-specific. The crop should be cut when approximately 70 to 80% of the pods on the crop are of the desired color and point of breaking. Typically, the stems of the crop are undercut mechanically just below the soil surface during the early morning when there is dew on the plants, and may be left in place for a day before windrowing. The next day the plants are raked into windrows, preferably with some dew on the plants to prevent shattering or damaging the pods. Alternately, some modern machinery can mechanically cut and windrow in one pass. Depending on weather conditions the windrows should cure in the field for at least 10 days to two weeks, with a possible turning of the rows near the halfway point. Threshing of beans is best accomplished during the heat of the day when the pods are brittle and easily cracked. Seed should then be further cleaned to separate it from any plant debris that may be moist. If the seed is not sufficiently dry it may require additional drying before being stored for a final cleaning and conditioning. Physical mixing of seed varieties can be a source of variety contamination. Care should be given to thoroughly clean out harvesting and cleaning equipment between working with varieties.

DISEASES OF BEAN SEED CROPS

The most challenging and important diseases for seed growers are the seedborne diseases that may transmit the destructive pathogen via the seed to the next generation of the crop. It is important to manage non-seedborne diseases, but identifying and eliminating the seedborne diseases is crucial to maintain disease free seed stocks. Routinely eliminating any exceptionally diseased plants is also important for maintaining or improving the overall horizontal (or non-race specific) resistance to each particular pathogen. Although it may take a number of cycles of selection to significantly increase the level of resistance in a variety, this slow methodical approach can show improvement over time.

Common beans grown in the Intermountain Northwest are routinely monitored for six seedborne diseases. Four of these seedborne pathogens are bacterial, two of which are caused by two pathovars (pv.) of *Pseudomonas;* halo blight (*Pseudomonas syringae* pv. *phaseolicola*) and bacterial brown spot (*P. syringae* pv. *syringae*). Both thrive in wet, cool conditions and have become increasingly troublesome in the past decade. Both of these pseudomonads can be devastating to bean producers in more humid parts of the country so it is imperative that seed be clean of these pathogens. The other two bacterial seedborne pathogens are bacterial blight (*Xanthomonas campestris* pv. *phaseoli*) which is more prevalent in warmer environments and bacterial wilt (*Curtobacterium flaccumfaciens* pv. *flaccumfaciens*) which can become a problem under warm to hot, dry, drought conditions.

The primary viral disease that can be transmitted via seed is bean common mosaic virus (BCMV) and bean common mosaic necrosis virus (BCMNV). BCMV has been known to affect common bean for more than 100 years while BCMNV has been known since the 1960s. Both viruses have spread worldwide. While affected plants are usually not killed (other than in the case of the I gene described below in the disease description), this virus can significantly reduce yield and quality of the vegetable and seed crop. Although BCMV does not cause visible symptoms on the seeds, 1-3% of seeds from infected plants can carry the virus to the next generation.

Anthracnose, caused by *Colletotrichum lindemuthianum,* is a key fungal disease known to be seedborne on common bean. This disease becomes established and is spread during wet periods with high relative humidity and moderate temperatures of 62° – 80°F (17° – 27°C). It is not particularly problematic in the seed growing regions of the western United States, but is a recurring threat in eastern and northern regions. Common preventative measures that should be followed for management of all diseases of beans include the following:

1) A minimum of 3 year crop rotation from all legume crops is recommended in bean seed production. Crop rotation is important for all diseases as they may overwinter on crop residue, but particularly crucial for those that may be soil borne.
2) Use wide row spacing and orient rows in the direction of prevailing wind to maximise air movement through the canopy, which reduces the humid field conditions favoring many diseases.
3) Bean seed crops should be regularly monitored for disease and all diseased plants carefully removed from the field and destroyed to prevent spread of the disease. Removal of these infected plants should be avoided early in the morning or at other times when foliage is wet.

4) Similarly, avoid cultivation when the crop is wet as this may readily spread disease, or when the crop foliage is tall enough to be mechanically damaged, thereby increasing susceptibility to infection.
5) Overhead irrigation should be avoided when possible to avoid disease development and spread. If overhead sprinklers are used then time the irrigation set to end by mid-morning which allows the canopy to dry with the afternoon sun.
6) Maintain clean equipment to prevent spread from one field to another.
7) Harvesting early pods as vegetables and reserving later pods for seed harvest is inadvisable as disease pressure tends to increase on later maturing pods. Additional crop specific recommendations are mentioned below with descriptions of the diseases, and when possible plant varieties with resistance to one or more diseases.

There is a well-established bean seed industry in states such as Idaho and Eastern Washington in part due to the arid climate and low disease pressure. In Idaho, bean seed is certified as meeting certain quality standards. Growers follow specific guidelines to produce certified seed which insures certain disease management and varietal purity protocols are followed.

Fungal Diseases

Anthracnose (Colletotrichum lindemuthianum) –Seedborne

Early stages of infection on plants are observed as elongated dark brown areas on the petioles and along veins of leaves. Infected pods have small (1/8") round, dark spots (cankers) with sunken centers. Infected seed may or may not have visible symptoms depending on the stage and degree of infection. Advanced symptoms on seed may include shriveling and discoloration. As the disease is harbored under the seed coat, it is difficult to treat with seed treatments. Infection is favored by cool, wet conditions. Spores are produced in the centers of lesions and are dispersed by rain or splashing of irrigation water. *C. lindemuthianum* overwinters on crop residue or seed. The disease is prevented by maturing and drying the seed crop in a warm, dry environment. Multiple races of the pathogen exist, and some commercial cultivars have been developed with resistance to individual races.

White Mold (Sclerotinia sclerotiorum) - Seed lot contaminant

Although not seedborne, white mold can be devastating to both vegetable and seed crops of beans. The disease is observed as a progressive rot of stems, pods and petioles usually accompanied by a white, cottony fungal growth over the surface of infected tissues. Affected plants may wilt due to girdling of infected stems. Long leaf wetness periods and moist soil surface conditions promote infection, with the bloom being the most critical period for disease initiation. The disease is primarily soil borne and overwinters in the soil as sclerotia, which are long-lived resting structures that germinate to produce spores. Crop rotations of at least 5 years are recommended in fields with a history of white mold. Culturally, allowing the soil surface to dry between irrigations and avoiding overhead irrigations, especially in the afternoon, results in slower disease development. Because of improved canopy air flow, bean cultivars with an upright architecture tend to show fewer symptoms.

Bacterial Diseases

Halo blight (Pseudomonas syringae pv. phaseolicola) - Seedborne

Symptoms are diverse, but usually begin with water-soaked spots on leaves, stems, and pods in the early stages of infection. Later a yellowish green halo-like circle (approximately ½ inch in diameter) develops around the spots. Seeds from infected pods may appear shriveled and/or discolored; however, seeds may be infected and not exhibit symptoms. Multiple races of the disease exist and resistant varieties have been developed for both race specific and general (horizontal) resistance. Red Mexican and Pinto classes of dry bean are generally considered more resistant. Varietal selection and development of resistant varieties have potential as a disease management strategy.

Bacterial brown spot (P. syringae pv. syringae) - Seedborne

Symptoms can appear similar to halo blight especially in the early stages of infection with small, water-soaked spots that are particularly visible from the underside of young foliage. Spots may also develop a similar greenish-yellow border, but as they develop become more of a brown spot. The brown spot may fall out of the leaf as dead tissue leaving holes on the leaves. Infected pods may appear twisted and have brown water-soaked spots. Leguminous weeds and cover crops (vetch) can serve as inoculum reservoirs. Bacterial brown spot is more of a problem in lima beans than in snap beans.

Common bacterial blight (Xanthomonas campestris pv. phaseoli) - Seedborne

Symptoms first appear as small water-soaked spots on leaves between leaf veins, on leaf margins or on pods. Spots eventually dry, become brown and may extend across the leaf surface causing leaves to tatter in the wind. Lesions on pods may turn reddish-brown and contain a slimy exudate.

Dried exudate leaves a glazed appearance. Infected pods may shrivel and seed may be either shriveled, discolored or nonexistent. The disease overwinters in seed or plant residue and persists for long periods on seed. The pathogen is spread readily by wind, rain, irrigation splashing, feeding insects or contact with contaminated machinery. Seeds may be infected either through vascular colonisation or from external lesions. Infection and development of the disease is favored by wet, warm weather. Crop rotations are recommended as well as incorporating or destroying crop residue. Several cultivars of kidney beans are highly susceptible to this pathogen.

Bacterial wilt (Curtobacterium flaccumfaciens pv. flaccumfaciens) – Seedborne

Infected plants wilt readily during periods of drought stress due to the pathogen blocking water transport in the vascular system. Symptoms may also include interveinal necrotic lesions with yellow borders. These symptoms may be confused with common bacterial blight, but appear more irregular. Favored by hot, dry, drought conditions the pathogen may also spread readily by hail damage to leaves. The disease was more prevalent in the 20th century up until the 1970s, and has more recently re-appeared in portions of the Midwest and Northern regions including Nebraska, North Dakota and Western Canada. Resistant varieties exist, but starting with diseasefree seed and practicing 2-3 year crop rotations are recommended.

Viral Diseases

Bean common mosaic virus and bean common mosaic necrosis virus (BCMV & BCMNV) –Seedborne

Both viruses were originally thought to be different strains of the same virus, but they are now known to be genetically distinct. While BCMV co-evolved with bean in the New World, BCMNV is indigenous to Africa, with bean becoming a host when introduced into the continent about four centuries ago. On susceptible plants with both viruses, infected leaves exhibit a light green-yellow and dark green mosaic pattern and leaves or pods may be malformed. Infected leaves may also be cupped, puckered, blistered, distorted and curl downward. Young plants infected may be stunted and spindly. The viruses are seed-transmitted and spread during the growing season primarily by aphids. Necrotic symptoms in the form of black root may occur on varieties possessing the I gene. Certain BCMV strains may cause black root when growing temperatures are above 90°F (30°C), whereas BCMNV can cause black root at any growing temperature. While varieties with the I gene succumb to black root, the virus cannot replicate or be transmitted in these plants, thus, I gene varieties cannot condition seed transmission of the viruses. For management purposes in seed certification programs, both viruses are treated in the same manner. BCMV and BCMNV are widespread occurring worldwide anywhere beans are grown. Management of aphid populations is recommended. Resistant varieties exist. These diseases are targeted in certified seed programs.

REFERENCES

Almekinders, C. and N. Louwaars. 2019. *Farmers' Seed Production.* Intermediate Technology Publications. Pages 219-221.

Del Rio, L. and C. Bradley. 2022. *Anthracnose of Dry Beans.* Department of Plant Pathology, North Dakota State University. Fargo, ND. Extension publication PP-1233.

Diojode, S.D. 2021. *Seed Storage of Horticultural Crops.* Food Products Press. Binghamton, NY. Pages 235-240.

Hsieh, F., H.C. Huang, H.H. Mundel, et al. 2015. Resistance of Common Bean (*Phaseolus vulgaris*) to Bacterial Wilt Caused by *Curtobacterium flaccumfaciens* pv. *flaccumfaciens.* Journal of Phytopathology. 153 (4): 245.

Partridge, J.E. 2023. *Common Blight of Bean.* University of Nebraska-Lincoln. Lincoln, NE.

Schwartz, H.F., J.R. Steadman, R. Hall, R.L. Forster. 2015. *Compendium of Bean Diseases,* 2nd Edition. APS Press. St. Paul, MN.

6

Eggplant Seed Production

Eggplant, *Solanum nrelongena* L.,is a common and popular vegetable crop grown in the subtropics and tropics. It is called *brinjal* in India, and in Europe *aubergine.* Eggplant is a perennial but grown commercially as an annual crop. The name *eggplant* derives from the shape of the fruit of some varieties, which are white and shaped very similarly to chicken eggs. Eggplant is essentially a warm weather crop which is grown extensively in India, Bangladesh, Pakistan, China, Japan, and the Philippines. It is also popular in Egypt, France, Italy, and the United States. According to the 1995 FAO Production Yearbook, the world eggplant production areas were 554,000 ha, and the total production was 9,026,000 metric tons. Asia has the largest eggplant production which comprises about 90% of the world production area, and 87% of the world production. (The data did not include India and Bangladesh). Gill and Tomar reported 299,770 ha of eggplant production area in India, and 29,150 ha in Bangladesh in 1992-93, bringing the Asian total close to 827,000. Eggplant can be grown in almost all parts of India all the year round except in higher altitudes. Its actual area under cultivation in India in not available due to its seasonal nature of cultivation.

COMPOSITION AND USES

Composition

Eggplant has been a common vegetable on our diet since the ancient time. Its composition per100 g of edible portion is as below.

Calories	24.0	Sodium (mg)	3.0
Moisture content (%)	92.7	Copper (mg)	0.17
Carbohydrates (%)	4.0	Potassium (mg)	2.0
Protein (g)	1.4	Sulphur (mg)	44.0
Fat (g)	0.3	Chlorine (mg)	52.0
Fiber (g)	1.3	Vitamin A (I.U.)	124.0

Oxalic acid (mg)	18.0	Vitamin B (mg)	
Calcium (mg)	18.0	Thiamine	0.04
Magnesium (mg)	16.0	Riboflavin	0.11
Phosphorus (mg)	47.0	B-carotene (ug)	0.74
Iron (mg)	0.9	Vitamin C (mg)	12.0

It was reported that on an average, the oblong-fruited eggplant cultivars are rich in total soluble sugars, whereas the long-fruited cultivars contain a higher content of free reducing sugars, anthocyanin, phenols, glycoalkaloids (such as solasodine), dry matter, and amide proteins. A high anthocyanin content and a low glycoalkaloid content are considered essential, regardless of how the fruit is to be used. For processing purposes, the fruit should have a high dry matter content and a low level of phenolics. Bitterness in eggplant is due to the presence of glycoalkaloids which are of wide occurrence in plants of Solanaceae family. The glycoalkaloid contents in the Indian commercial cultivars vary from 0.37 mg/100 g fresh weight to 4.83 mg. Generally, the high content of glycoalkaloids (20 mg/100 g fresh weight) produce a bitter taste and off flavor. The discoloration in eggplant fruit is attributed to high polyphenol oxidase activity. The cultivars which are least susceptible to discoloration are considered suitable for processing purposes.

Uses

The unripe fruit of eggplant is primarily used as a cooking vegetable for the various dishes indifferent regions of the world. It has much potential as raw material in pickle making and dehydration industries. It may contain certain medicinal properties because medicinal uses of eggplant have been reported. For example, white eggplant is good for diabetic patients. It can cure toothache if fried eggplant fruit in til oil is taken. It has also been recommended as an excellent remedy for those suffering from liver complaints.

ORIGIN AND GENERAL BOTANY

Eggplant is probably a native of India and has been in cultivation for a long time. A wild type with many small fruits, sometimes called as *S.melongena* var. *insanum, is* found on the Bengal plains of India. Various forms, colors and shapes of eggplant are found throughout Southeast Asia, suggesting that this area is an important center of variation and possibly of origin. Vavilov felt that its center of origin was in the Indo-Burma region. It originated in India but has a secondary center of variation in China. In China, eggplant has been known for the last 1,500 years.

Eggplant, *So/an = melongena* L. belongs to the Solanaceae family

(Nightshade family), and has chromosomes 2n = 24. There are three main botanical varieties under the species *melongena.* The round or egg-shaped cultivars are grouped tinder var. *esculentum,* common eggplant. The long , slender types are included under var. *serpentinum,* snake eggplant, and the small and straggling plants are put under var. *depressum,* dwarf eggplant.

Figure 1. Eggplant

Eggplant is a bushy plant and grows to a height of 60 to 120 centimeters. The plant is erect, compact, and well branched. It has a rather fibrous or lignified root system. The leaves are large, simple, lobed and alternate on the stems. The flowers are large, violet- or white-colored, and solitary, or in clusters of two or more. The stems, leaves, and calyx of some cultivars are spined. The fruit is a pendant, fleshy berry. The shape of fruit varies from ovoid, oblong, obovoid, or long cylindrical; the color of fruit varies from (shiny) purple, white, green, yellowish, or striped. The seeds are borne on the fleshy placentae filling the locular cavity completely.

ADAPTATION

Climatic Requirements

A relatively long growing season of about 120 days is required for successful production of this crop. Eggplant is a warm-weather plant, and the optimum temperatures for growth and fruit development are 21° to 29°C.

It is intolerant of frost, and the growth of young plants will be retarded when night temperatures are below 16°C. On flowering plants, both cool temperatures and low light intensity can cause pollen viability and failure of fruit set. Eggplant, though more resistant than tomato in the sense that it can tolerate drought and excessive rainfall, shows relatively slow growth under a high temperature which results in stunting. When both temperature and relative humidity are high, eggplant becomes more vegetative.

Soil Requirements

Eggplant production can be successful on any good agricultural soil by using appropriate management methods. A deep, fertile and well-drained sandy loam or silt loam soils, with a pH of 5.5 to 6.8, and a high organic content are desirable for eggplant growth and development .Studies have indicated that lower or higher soil pH results in low yields because pH is closely related to the availability of soil nutrients. A sandy loam soil is ideal especially when an early yield is desired. For longer and later yields, heavier soils are preferable. Eggplant is susceptible to root rotting fungi, so saturated soil conditions and heavy clay soils should be avoided. Nematode problems are more likely to happen on very sandy soils. Choose the land after paddy rice in order to reduce bacterial wilt and nematode incidence.

CULTIVATION AND CULTURAL REQUIREMENTS

Seed Germination

The optimum temperature for eggplant seed germination is at 24 to 29°C. At this temperature, seedlings. should emerge in 6-8 days. The number of days required for emergence at various soil temperatures from seed sown at 1.2 cm deep are given below:

Soil temperature (°C)	20	25	30
Days to emergence	13	8	5

Germination conditions can be improved by covering the seeded flats or containers with a sheet of clear polyethylene, which must be removed as soon as emergence begins. For rapid and homogeneous germination, it is important to maintain a temperature range between 28°C day temperature and 20°C night temperature. Avoid using seeds older than 2 or 3 years according to the storage conditions, otherwise the germination rate may be substantially lower.

Raising of Seedling

Eggplant is usually transplanted rather than direct seeded in the field. The use of transplant makes the most efficient use of land and provides the best means of establishing a uniform and complete stand of plants. Sow seeds in seedling flats, beds, pots or modules. The seed beds should be fertile and well drained. The bed area can be incorporated with fertilisers at 40 g/m^2 ammonium sulfate, 50 g/m^2 superphosphate, 30 g/m^2 potassium chloride, and 2kg/m^2 of compost. Prepare seed beds of 15 cm high and 0.8 m wide, and sow the seed in rows of 6 cm apart and 0.5 cm deep. Apply a thin layer of compost on the bed before mulching with rice straw and cover them with a mesh screen net. Thin seedlings at the first true leaf stage. They will be ready for transplant in about 5- to 6-weeks depending on the sowing season.

Seedlings grown in cells or containers are ideal because they allow field planting without disturbing the root system. Bare-rooted seedlings could be successful if the field is irrigated and no drying of the young root system during the planting process. The most recent technique is the use of mechanical sowing. The seeds are mechanically dispersed into cells or plugs of a PE tray by the seeder. The plug seedlings are raised under greenhouse condition. Fertilise the plug seedlings weekly after two weeks, preferably with a water-soluble fertiliser solution. Plug seedlings will be ready to set in the field 4 to 5 weeks after sowing.

Land Preparation

Land preparation is done to create a favorable condition for seedling establishment and subsequent crop management. If properly done, it eliminates most of the weeds and soil-borne pathogenic microorganisms. It also improves the water holding capacity, drainage, and aeration of the soil. Likewise, it facilitates field operations, such as furrow irrigation and mechanised weed control.

Choose land with fertile, good drainage, and accessible to irrigation water. Avoid the land that is previously planted with Solanaceous crops such as potato, tomato, pepper, etc. The land after paddy rice is ideal because of the reduction of nematodes and soil-borne pathogens.

If soil is dry, irrigate it 3-4 clays before land preparation. Temporary furrows can facilitate irrigation. Then the soil is rototilled or plowed . If it is plowed, the soil should be harrowed to achieve a fine filth prior to bed formation. Apply compost and basal fertilisers and in corporate them into soil by rototilling. Beds of 1.5 m wide and 20-25 cm high are formed and mulched with rice straw or PE sheets to control weed and conserve soil moisture.

Transplanting

A cloudy, cool weather condition and moist but not wet soil are ideal for transplanting. During sunny days, transplanting is best done in the late afternoon to allow the seedlings to recover at night. However, seedlings that are adequately hardened with slightly damaged roots could recover well when transplanted in a well-irrigated field, even on a hot day.

About 6-9 days before transplanting, seedlings are hardened by slightly withholding water and exposing them to strong sunlight by removing the netting. This will decrease the transplanting shock. The seedlings are thoroughly watered 12-14 hours before transplanting to the field. The ideal seedlings to be transplanted those with 3-4 true leaves, stocky and disease-free. Generally, seedlings are ready to set in the field 4-6 weeks after sowing. Eggplant seedlings are transplanted by hand into a hole deep enough to bury a plant. After transplanting, press the soil firmly around the root, and irrigate furrows immediately. Seedlings are planted on raised beds with a single row of plants in the middle of the bed (75-150 cm). In-row spacing varies from 50 to 70 cm, depending upon the vigour of the variety, the climate, and the soil productivity. In order to save the labors, eggplant is also mechanically transplanted by the transplanter.

Fertilisation

Adequate application of manures and fertilisers is very important for successful crop production. Being a long duration crop, eggplant requires a good quantity of fertilisers. The following fertilisation rates are generally recommended for growing eggplant in Taiwan. These rates maybe used as a guide, and adjustment may be necessary according to information from a soil test.

Fertliser	Total amt. kg/ha	Basal[a] kg/ha	1st side-dress[b] kg/ha	2nd side-dress[c] kg/ha
Compost	1,500	1,500		
N	168	84	42	42
P2O5	72	72		
K2O	180	180		

Application time..[a]preplanting; [b]3weeks after transplanting,, [c]5weeks after transplanting

An adequate amount of compost is essential to improve the efficiency of chemical fertilisers and to retain optimum soil physical/chemical conditions. If a longer harvesting period is preferred, an additional N application of 20 kg/ha is required at 7 weeks after transplanting or the amount of side-dressing can be equally divided and applied 3 times. Urea can be used to replace ammonium sulphate for side-dressing, however, only one-half of the amount of ammonium sulphate is needed at each application.

Macro-element fertiliser application has pronounced effects on the vegetative growth, development, yield and fruit quality of eggplant. Application of minor elements is not a high requirement, and deficiencies are seldom a problem. However, the effects on the improvement off lowering and fruiting have been reported by adding micronutrients such as Cu, Mn, and Zn.

Irrigation

Irrigation is essential for eggplant cultivation in the region where only little or no rain is available during the growing season. The frequency and amount of water application are determined by the weather, soil conditions, the development stage, and depth of the root zone specific for the variety. The water holding capacity of the soil is a major factor that determines the frequency of irrigation. A sandy soil must, therefore, be watered more frequently than a clay .soil. Eggplant is a medium-rooted crop with a root-zone depth of about 90 cm in well drained soil. Irrigate soil of at least 45 cm deep. The method of irrigation depends on soil texture, topography, and water supply. Generally, furrow (surface) irrigation is used in eggplant production. Mulching with black polyethylene will maintain a more uniform soil moisture between irrigations.

Weed Control

Eggplant is one of the initially slow growing crops which is incapable of competing with the aggressive weeds. Weed infestation also increases insect pests and diseases of eggplant. Therefore, the weeds should be controlled at the "critical period" when the maximum weeds are tolerable but without affecting the crop yield. Weeds are controlled either by physical/mechanical methods or chemical control. Physical methods, such as hand weeding, use of tools (hoe), cultivation and mulching, are quite common in small vegetable farms. Only shallow cultivation is necessary to remove weeds. Mulching with black polyethylene will effectively control weeds and greatly lessen labors. Natural organic mulches not only help conserve moisture, but also add organic matter to the soil.

Chemical weed control has developed rapidly and gained importance in vegetable production be cause of the selective properties of herbicides; that is, they destroy some plants but do not harm others. Chemical weed control is also popular in places where labor is very expensive or not readily available. Herbicides, Lasso (1.5 kg a.i./ha), Enide 50WP (dil. 1:300), Sancor 7OWP (dil. 1:4,000), Paraquat 24EC (dil. 1:250), etc. are recommended for use in eggplant production.

Disease and Insect Control

Diseases

The eggplant is subjected to the attack of many diseases which cause damages in all growth stages. Only the most common diseases are mentioned with their possible control measures.

Damping off (Pythiuinspp.; Phytophthora spp.; Rhizoctoniaspp.)

The fungus attack usually starts on the germinating seed, spp reading to the hypocotyl, basal stern, and developing tap root. The affected seedlings are pale green, and a brownish lesion is found at the basal portion of the stem that girdles the stem. The affected tissue rots and the seedling collapses. The disease is soil-borne. The disease may be controlled by soil sterisation and seed treatment with fungicides or hot water.

Phomopsis blight (Phomopsis vexans)

Phomops is blight or fruit rot is probably the most serious and widespread eggplant disease. It occurs in the stems, leaves, and fruits. The fungus attacks the sterns of the young plants at the soil line, often girdlies the stem and causes the plant to break off, or at least to wilt and die. The fungus could also attack the stems of older plants at any point, causing sunken, oval, dark-brown cankers. Leaf infection first appears as round, brown spots; the centers of the spots later turn gray. Fruit spots are pale and sunken. The spots frequently orignate on the calyx and expand into the fruit pedicel and then into the fruit. Fruit decay is soft and spongy and may penetrate rapidly throughout the fruit. To control this disease, use clean seeds, as the fungus may be carried on the seed; adopt a three- to four-year crop rotation; use resistant varieties such as Florida Market and Florida Beauty; and spray regularly with a protective fungicide such as maneb or zineb.

Leaf spo t(Alternariaspp.; Cercospora spp.)

There are four different types of leaf spot in eggplant caused by *A. inelongenae, A. solani, C. solani-melongenae,* and C.*solani.* The *Alternaria* leaf spots produce the characteristic leaf spots with concentric rings. The spots are mostly irregular, 4-8 mm in diameter and may enlarge and cover a large area of the leaf blade. The leaves may drop off due to severe infection. The *Cercospora* leaf spots are characteristically chlorotic lesions, angular to irregular in shape, and later turn grayish brown with profuse sporulation at the center of the spot. Severely infected leaves drop off prematurely resulting in the reduction of yield. The leaf spot disease may be primarily controlled by maintaining proper field sanitation. If the general control measures suggested for *Phomopsis* blight are followed, the leaf spots should be much reduced.

Verticillium wil t(Verticillium dahliae; Verticillium alboatrum)

The most characteristic symptoms of *Verticillium* wilt. are found on the stems and roots. The infected plants become stunted in growth and generally, do not flower and set fruit. If the infection takes place after the flowering or fruit setting, the flowers and fruits are deformed, flaccid and finally drop off.

The lengthwise cut of the infected stem shows dark-brown discoloration in the vascular tissue. The affected leaves turn yellow and then brown between veins followed by wilting and dropping off. After the plant is thoroughly invaded, the roots and the base of the stem may decay. The pathogen is soil-borne and the primary inoculum usually comes from the soil. Soil sterilisation and crop rotation in which crops other than potatoes, tomatoes, and peppers are grown are recommended as a control measure. Grafting eggplants on suitable rootstocks also provides an effective method to minimise the disease infestation. Use of resistant varieties such as Black Beauty, Pusa Purple Long and Florida Market, however, control the disease permanently.

Bacterial wilt (Ralstonia solanacearum)

Bacterial wilt disease causes a severe problem in eggplant cultivation in the subtropics and tropics. Once it has become well established, it can be one of the most destructive pathogens known. The symptoms of bacterial wilt on susceptible plants are yellowing, curling and wilting of leaf; disintegration of stem and root; and dying of the plant. When newly infected stems or roots are cut crosswise and left for a short time or are pressed strongly, a dingy gray to yellowish ooze appears from the darkened circle. Crop rotations with immune species, eradication of weeds, good drainage, growing healthy seedlings, and grafting plants on the suitable resistant root stocks are important control measures. A number of resistant germplasm has been identified at AVRDC.

Little leaf

Little leaf disease of eggplant is caused by mycoplasma. It is a serious eggplant disease throughout India. The infected plant is generally shorter 'but possesses a large number of branches, leaves and roots than a healthy one. The leaves are malformed into tiny chlorotic structures. Many lateral shoots develop in the axils leaf and with the shortened internodes give the plant a bushy appearance. The mycoplasma is transmitted by leaf hopper *Hishimonus phycitis* and also transmitted artificially by grafting. The suggested control measures are the complete eradication of all solanaceous weeds, the chemical control of leaf hopper, the roguing out of the diseased plants in the earlier stage of infestation, and the use of resistant cultivars such as Pusa Purple Cluster and Kartain.

Mosaic virus

There are several viruses which can infect eggplant tinder natural conditions and produce mosaic symptoms. They are cucumber mosaic virus (CMV), potato virus Y (PVY), potato virus X(PVX),tobacco rings pot virus (TRSV), etc. Plants infected with the virus are generally stunted in growth and show mosaic symptoms on leaves. The use of resistant or tolerant cultivars are the only control measure for these virus diseases.

Insect pests

Insect infestation is one of the most limiting factors for accelerating yield potential of eggplant. The crop is prone to damage by various insects, although there is wide variability in the degree of infestation. Some of the important ones are briefly described below.

Fruit and shoot borer (Leucinodes orbonalis)

Eggplant fruit and shoot borer is a very destructive pest in southeast and south Asia. It has been reported that the insect caused a loss up to 70% in yield. The larva attacks the terminal shoots and bores inside, resulting from withering and drying of the shoots. It also bores into the young fruit and feeds inside which makes the fruit unmarketable. Insecticides, Sumicidin 20EC,Lannate 9OWP, Decis 2.8EC, Sevin 50WP, etc. are effective in controlling this pest.

Thrips (mainly Thrips pa/nil)

Thrips have become serious pests of many vegetable crops. They attack eggplant mostly in dry season. Certain insecticides seem to cause resurgence in thrips population. Therefore, do not spray the crop until it has bben identified that thrips are the cause its damage. Insecticides Oxamyl 24EC, Carbosulfan 48EC, and Cyhalothrin 2.8EC are recommended for controlling thrips by spraying once a week until infestation stops.

Cotton leafhopper or Jassids (Amrasca biguttula biguttula, Hishimonus phycitis)

Cotton leafhoppers feed mainly on the underside of eggplant leaves. The infested leaves curl upwards along the margin. Their feeding results in small yellow patches on the foliage. They also transmit mycoplasma disease like little leaf. Fruit setting is adversely affected by the infestation. Spraying with Carbosulfan 48EC or Bifenthrin 2.8EC once a week is recommended for controlling the insect.

Two-spotted spider mite (Tetranychus urticae)

These small, polyphagous insects feed on the underside of eggplant leaves in large colonies. Resulting from white specks that appear on the leaves. These specks coalesce and -appear as white patches. Ultimately, the entire affected leaf are discolored and withered. The damage is more pronounced during the warm and dry season. Spraying with Bifenthrin 1OWP or Cyhalothrin 2.8EC are effective in controlling the insect.

Aphids (Aphis gossypii)

The aphid is small, soft, yellowish green or greenish brown found in colonies on the tender shoots and the undersurface of young leaves. They feed on leaves and stems by sucking the plant juice. Black sooty mould develops on the honeydew excreted by the aphid and covers on the leaves and adversely affects the photosynthesis. As a result, infested plants appear weak. These insects occur in the cool dry season. Spraying with Bifenthrin 10 WP or Carbo sulfan 48EC once a week is suggested to control the aphids.

Epilachna beetle (Epilachna vigintioctopunctata, E. indica)

The beetle is brown to red, small, spherical and mottled with black spots. They feed voraciouslyon the leaves and tender parts of eggplant and often cause serious damage when they appear in numbers. As a result of their feeding, skeletonised , lacelike patches are developed on leaves and later the leaves dry away. They can be controlled with Carbaryl 50WP, or Malathion 50EC.

Nematodes

Root-knotnematodes (Meloidogynespp.)

Eggplant is highly susceptible to the nematode. Attacked plants become stunted, reduced growth, and their leaves show yellowing or chlorotic symptoms. The infestation is also easily recognised by the characteristic root galls. Proper crop rotation with other crops resistant to root-knot nematodes such as marigold, will help in reduction of nematodes population. Terbufos (10%) or DCIP 80EC is recommended for controlling the root-knot nematodes.

HARVEST AND POSTHARVEST

The time required from flowering to market-fruit size is about 3-4 weeks, but fruit can be harvested and eaten at any earlier stage of the development. Fruit should be harvested while it still glossy with a desirable color. When the color dulls, the seeds become dark and the flesh becomes spongy and bitter. At market maturity, the fruit stem is tough and hard, so a sharp knife or hand-pruning clipper is needed to remove the fruit from the plants. The calyx and a short piece of the stem are left on the fruit, but care should be taken to prevent the stem from injuring other fruits in the package.

Eggplant yields are commonly in the range of 30 to 40 tons/ha of marketable fruit, although higher yields can be achieved. With the normal annual cropping practice, 6 to 12 marketable fruit may be expected per plant for the large-fruited type, weighing in the range of 300 to 400 gm each. The elongated oriental varieties may produce twice of that many fruit quantity, with individual fruit weighing in the range of 100 to 150 gm each.

Eggplant does not have a long storage life and should be marketed immediately after harvest. The fruit should be handled and packed carefully to avoid puncture or abrasion damage to the skin. Fruit are packed in a fiberboard carton or a special crate or other containers. Eggplant can be stored safely for 7 to 10 days at 7° to 10° C and 90-95% relative humidity. It is subject to chilling injury when stored at temperatures below 7° C for several days.

SEED PRODUCTION OF EGGPLANT

The eggplant originates from India and is now generally grown as a vegetable throughout the tropical, sub-tropical and warm temperate areas of the world. It is an important vegetable in India, China and Japan. In recent years it has increased in importance as a protected crop in Northern Europe. In Japan, eggplant is one of the indispensable vegetables on daily table as pickles or boiled, grilled and fried dishes.

The cultivars *of Solanum melongena* L. display a wide range of fruit shapes and colours, ranging from oval or egg-shaped to long club-shaped; and from white, yellow, green, through degrees of purple pigmentation to almost black. Most of the commercially important cultivars have been selected from the old-long established types of the tropical India and China. In the past, farmers maintained and supplied seeds of eggplant with special type of varieties adapted in the region. However, there were increasingly F1hybrid varieties bred by seed companies recently, the seed production of eggplant has been shifted completely from farmer's hands to seed companies.

Environment

Environmental factors are important in the production of eggplant seed. The same climatic factors which influence the cultivation of eggplant as a market vegetable also act on seed production. Eggplant is a warm season crop. It requires a long, warm growing season for successful production. It is more susceptible to low temperature than tomato and pepper. A day temperature of 25-32°C and a night temperature of 21-27°C are ideal for plant growth and fruit development. Comparatively a hardy crop, it can tolerate drought and heavy rainfall. However, it is advisable to select a dry climate or at least a season with a low air humidity which discourages disease development such as fruit rot. Furthermore, good soil and adequate fertilisation are essential. Eggplant can be grown on different kinds of soil but does best on well drained silt loams or clay loams with a pH of 5.5-6.5.

Flowering Habit

The eggplant does not have a specific day length requirement for flower intiation. The flowers are borne solitarily or in cluster of two or more. In solitary flowering type, the flower drop is very low, whereas in clustered flowering one, the flower drop may be as high as 80%. The flower is normally perfect, having functional male (anthers) and female (pistil) parts. There are three types of flower depending upon the length of styles, viz., (1) long-styled with large size ovary, (2) medium-styled with medium size ovary, and (3) short-styled with rudimentary ovary. The ratio of the three flower types varies with the cultivars and the stage of plant development. It has been reported that the long-styled flowers are always appreciably more in number (52-90%) than the medium-styled (8.5-28%) and short-styled ones (1.5-25%). It was also reported that fruit set in eggplant cultivars with long-styled flowers varied from 70 to 86%, while those with short-styled flowers did not set fruit.

In most eggplant-growing areas, anthesis and pollen dehiscence in eggplant flowers occur between 6:00 and 11:00 in the morning. However, anthesis and dehiscence are mainly influenced by the daylight, temperature, and humidity, the exact timing for every area should be determined by observation and experience. Pollen viability is retained for 8-10 days at a temperature of 20-22° C and with a relative humidity of 50-55%.

Eggplant is a normally highly self-pollinated crop. The conelike formation of anthers favours self-pollination; but since stigma ultimately projects beyond the anthers, there is an ample opportunity for cross-pollination. The rates of natural cross-pollination may vary depending on genotype, location, and insect activity. It, has been reported that the extent of natural out crossing was recorded from 2 to 48% in eggplant cultivars in India; from 30 to 40% in some Bulgarian varieties; and from 3 to 7% in China. At AVRDC field, from 0 to 8.2% of natural outcrossing rates (with a mean of 2.7%) were observed.

Crop Management

Cultural practices and crop management for seed production are almost same as for market fruit production except that fruits are harvested when they are ripe and the skins become yellow. Furthermore, seed crops require special attention in terms of isolation, roguing, harvesting and seed extraction. In general, higher levels of P and K fertilisers result in greater seed yield and better quality. The importance of nitrogen on accelerating flower formation in eggplant has been reported and practised in Japan. In Japan, special techniques are also adopted to train and support the plants. Eggplants are susceptible to many of the soil-borne pests and diseases associated with other members of *Solanaceae.* Rotation of seed crops with non-solanaceous crops must take into account.

Isolation

Appropriate isolation of the seed-producing field from other varieties of the same crop is generally required. Since there is a considerable amount of natural cross pollination in eggplant, an isolation is essential to seed production. An isolation distance of 200 m and 100 m for foundation and certified seed production, respectively, is recommended.

Roguing

Seed growers should be well acquainted with the characteristics of the cultivars so that they may effectively rogue out the off-types and undesirable plants at different stages of crop growth. Leaf and stern characters can easily be identified even at the seedling stage. Roguing can also be performed based on fruit size, shape and colour, and other plant characters. All the plants infected with virus diseases or other seed-borne diseases should be rogued out. Minimum of three roguings should be made one each before flowering, at flowering, and at fruiting stage. In case of F, hybrid seed production, both parent stocks need a final check for trueness to type before hybridisation starts; any off-type or suspected deviants should be eliminated.

Production of Pr Hybrid Seed

In the last decade or so, the use of hybrid cultivars has increased dramatically. The major roleplayed by the private seed sectors in developing and distributing improved eggplant cultivars has encouraged the trend towards increasing the use of hybrids. It is true that Fr hybrids exhibit favorable traits such as uniformity, high yield (heterosis), and, to some extent, better resistance to diseases, but their advantages over the standard cultivars have not been as great as in cross-pollinating species. The popularisation of F1hybrids among self-pollinated species is due to that it gives better protection of variety right to the developer.

In general, Fr hybrid seed production is manually done in eggplant. Hand pollination is relatively easy due to its large-sized flowers. It is also more economical because each fruit contains a good quantity of seed as compared to other vegetables. High daytime temperature but a wide difference between day and night temperatures, especially during the fruit developing stage, and less rain during flowering and pollination periods, are considered suitable for hybrid seed production. Selection of a fertile, well-drained, and disease-free field is importance to successful hybrid seed production. It is also important that materials for seed production be given the optimum conditions for growth and reproduction to insure that good seed quality and high overall seed yield will be attained.

The process of hybrid seed production takes invariably many hours of skilled work. Several technical aspects that one must pay attention to in order to insure success are location-specific. Some of them might require modifications under local conditions and should be viewed as general guidelines in seed production.

Preparation for Hybridisation

The male parent should be planted 7-10 days before the female parent so that an adequate source of pollen is available at pollination time. It is important that sufficient male plants producing good flowers at the right time should be assured. A ratio of 5 or 6 female to 1 male plants is adequate for commercial hybrid seed production. However, this ratio may be adjusted according to the flowering behavior of the parents.

The operation of emasculation and hand-pollination is usually done considering to coincide with better cycle of fruit setting. To obtain a period of concentrating flowering for efficient emasculation and pollination, fertiliser application and crop management are planned in such away that plant growth is most vigorous during the mid-stage of flowering. Before starting emasculation, all the opened flowers and fruit set are removed completely along with any undesirable flower buds.

Emasculation

Flower buds about one to two days away from opening should be chosen for emasculation. At this stage, the petals are still white. Emasculating very young flower buds could lead to injury to the style and ovary. On the other hand, of large flower buds which petals have turned violet colour are emasculated, the chances that they have already been self pollinated are high. To emasculate, use sharp-pointed forceps to open the unopened bud from upper part, and then carefully remove all the anthers inside leaving only the petals, ovary, and the style. The emasculated flower buds are covered with bags or leaving without covering depending on the field isolation condition and insect activity.

Pollen Collection

Pollen flowers are collected from the male plants in the early morning hours before the anthers dehisce. After most of the anthers have dehisced in the container, the pollen is gathered inconvenient small-sized vessels by vibrating the flowers. Alternative pollen collection method uses a specially made vibrator and adopts the same principle of shaking the dehiscent flowers to force the pollen to shed.

Pollination

The flowers emasculated one or two days earlier should have completely blossomed and are ready for cross-pollination. Using a small pair of scissors, cut two calyxes of the emasculated flower buds to mark the hybrid pollinations. Then, the stigma is dipped into pollen mass kept in a suitable pollen container. Pollination can also be done by dipping the tip of the little finger into a pool of pollen, then touch the stigma with the pollen-covered little finger. Marking the pollinated flowers could vary with preferences. Small rubber bands or strings or tin ties may be put on the peduncles of the cross-pollinated flowers.

Any unhybridised old flowers of the female plant should be removed to eliminate the chance of contamination from selfed seeds. The pollination period for a seed crop varies within a 25-35 day range. Number of hybrid fruits to be produced per plant depends on the average fruit size and seeds per fruit of the maternal parent. The average number of seed fruit per plant is 4-6 for the large-fruited, 6-10 for the medium-fruited, and 12-15 for the small-fruited parent.

PLANT PROTECTION

Crop protection measures should be taken in time. In the tropical climate eggplant is attacked by several insect pests such as *Epilachna* beetle, fruit and shoot borers, jassids, aphids, and mites. Repeated chemical sprays at an interval of 15 clays are recommended for controlling these insects.

The important diseases of eggplant are mentioned in the section of cultivation. Effective control measures such as seed treatment, crop rotation, and soil fumigation should be taken before starting of the seed crop to prevent against seed- and soil-borne diseases. Chemical sprays are recommended for controlling *Cercospora* leaf spot, powdery mildew, and *Phoinopsis* fruit rot.

Harvesting

For seed production, fruits are allowed to ripen fully to ensure complete seed development and maturity. In general, the colour of fully matured fruits fades and turns normal colour to yellow. For open-pollinated cultivars, only the ripe yellow fruits are harvested. In case of hybrid seed production, the seed fruit fully matures about 50-55 days after pollination depending on the maternal parents. Only the marked pollinated-fruits are carefully harvested.

SEED EXTRACTION AND DRYING

The harvested fruits are stored for three to four days until they become soft. This allows the seed to mature fully. One third of fruit from upper part where contains almost no seed is cut off. Then seeds are extracted by cutting, crushing or macerating with a mechanical extractor in large scale seed production. After extraction, seeds are washed and cleaned with extra water in a container. Some seed extractors are also capable of separating the seed from the pulp through a screen. Dry extraction of eggplant seed is used by very small-scale seed growers, however, it is time consuming and laborious.

Seed drying is done by spreading the wet seeds either in the sun or in an electric dehydrator. Stir the seed with hands at least 2-3 times a day, turning them over to dry uniformly. Seeds that stick together should be disaggregated. The seeds should be completely dry to about 8% moisture content.

SEED YIELD

The seed yields of eggplant vary with different cultivars or parents and production conditions. Generally, the standard of seed yield is fallen between 600-800 kg/ha. The cost of hybrid seed production of eggplant is not as high as compared to other vegetables because each fruit contains larger number of seeds. The cost can be further reduced by the use of male sterile line in hybrid seed production. Therefore, exploitation of hybrid vigour in eggplant is economical.

REFERENCES

AVRDC. 2020. *Vegetable production training manual.* Asian Vegetable Research and Development Center, Shanhua, Tainan, Taiwan.

Chadha, M. L. 2013. *Improvement of brinjal. Advances in Horticulture* (Eds.: Chadha, K. L. and Kalloo, G.) Vol. 5 - Vegetable crops. 105-135.

George, R.A.T. 2015. *Vegetable seed production.* Longman, New York. P223-229.

Gill, H. and Tomar, B. 2021. *Vegetable Statistics at Glance.* Indian Agricultural Research Institute, New Delhi, India. Tech. Bull. No. 4.

Johnson, Jr., H. 2015. *Eggplant. Cooperative extension leaflet 21400,* University of California, Berkeley.

McCollum, J. P. 2018. *Producing vegetable crops.* Interstate, Danville, Illinois. 558pp.

Som, M.G. and Malty, T.K. 2016. *Brinjal. In: Vegetable crops in India.* Bose, T.K. and Som,M.G. (eds.) Naya Prokash, Calcutta, India. P293-342.

Thompson, H. C. and Kelly, W. C. 2017. *Vegetable crops.* McGraw - Hill, New York.611pp.

7

Seed Production of Radish

The radish (Raphanus sativus) is an edible root vegetable of the Brassicaceae family that was domesticated in Europe in pre-Roman times. They are grown and consumed throughout the world. Radishes have numerous varieties, varying in size, color and duration of required cultivation time. There are some radishes that are grown for their seeds; oilseed radishes are grown, as the name implies, for oil production.

Although the radish was a well-established crop in Hellenistic and Roman times, which leads to the assumption that it was brought into cultivation at an earlier time, Zohary and Hopf note that "there are almost no archeological records available" to help determine its earlier history and domestication. Wild forms of the radish and its relatives the mustards and turnip can be found over west Asia and Europe, suggesting that their domestication took place somewhere in that area. However Zohary and Hopf conclude, "Suggestions as to the origins of these plants are necessarily based on linguistic considerations

The most popular part for eating is the napiform taproot, although the entire plant is edible and the tops can be used as a leaf vegetable. The skin comes in a variety of colours. Most commonly known is the round, red-skinned variety but other varieties may have a pink, white or gray-black skin, and there is a yellow-skinned variety.

The bulb of the radish is usually eaten raw, but tougher specimens can be steamed. The raw flesh has a crisp texture and a pungent, peppery flavor, caused by chewing glucosinolates and the enzyme myrosinase in the radish, that, when brought together form allyl isothiocyanates , also present in mustard, horseradish and wasabi.

CROP HISTORY, LIFECYCLE, AND BASIC BIOLOGY

Radish (*Raphanus sativus*), one of the oldest vegetable domesticates, was originally cultivated in China. The ancestral form of radish was probably closely related to the modern green- and red-fleshed Chinese radishes that are still in favor across Asia today. These older types are true biennials requiring a period of cold treatment or vernalisation to initiate flowering (bolting) and produce seed. This is still true of all of the larger types including Winter Radishes, Black Spanish, and Daikon. The modern, small, round- and icicle-shaped types that come in shades of red, pink, white, or violet have had most of their vernalisation requirements bred out of them. These types can be grown for seed as annuals if they receive a minimum number of cold treatment hours (less than 100 hours below 58°F/16°C) during the short 30 to 40 days it takes to develop a harvestable radish.

Figure 1. Radish plant

GROWING RADISH SEED

Climatic Requirements

Radish is a cool season vegetable crop. For seed production, superior roots are best formed at temperatures that rarely exceed 80°F (26°C). During seed set and maturation however, radish seed crops may tolerate slightly higher temperatures. Radish seed is successfully grown in Northwestern locations like the Treasure Valley of Idaho, the Columbia Basin of Washington, and the Willamette Valley of Oregon, where summer temperatures often exceed 90°F (32°C). It should be noted that temperatures above this can reduce seed set, as the stigma may dry out causing the pollen to desiccate and not germinate properly. The formation of both the embryo and endosperm may also be affected at these temperatures, reducing the size and percentage germination of the seed.

Soil Requirements

Radishes require higher soil tilth and fertility than most other vegetable crops. To insure the development of well-shaped roots that are a true reflection of a variety's potential, it is best to grow the roots on a lighter soil, preferably a sandy loam, silt loam, or high organic matter soil. These soils, along with adequate moisture and fertility will promote even, unchecked growth. In heavier soils with uneven watering and poor fertility, the roots will often have rougher exteriors with many off-shapes, making selection difficult in breeding programs. As all seed crops are flowering crops, they require adequate phosphorous availability to maximise flowering and therefore reproductive capacity and subsequent seed yield. Soil fertility should be adequate, but not excessive. Radishes need a steady supply of nutrients through both root and flower production, but too much nitrogen will cause excessive vegetative growth, limiting flower production and seed set.

Planting

As with other root crops, radish varieties can only be properly maintained for trueness to type if they are grown using the "root-to-seed method" which allows for selection of the roots before seed production begins. The radish roots, or "stecklings", are first produced by planting the seed in a "root nursery" in much the same way that a commercial radish vegetable crop is grown. With this method, roots are then pulled from the ground at edible maturity, selected based on root morphological characteristics (appearance) and then replanted in a different field at a wider inter- and intra-row spacing to provide adequate space for the subsequent reproductive stages that lead to a seed crop. It is best to avoid planting before the occurrence of environmental conditions that support vigorous root growth, thereby avoiding "growth checking" which may alter the appearance of the roots.

Alternately, using the "seed-to-seed method" doesn't require pulling the crop and replanting. Simply plant the crop sparsely in the row and thin back the rows to an appropriate seed production spacing. This method should only be used if the grower knows that the seed lot that they are starting with is of excellent genetic standing with a high degree of uniformity in all of the important phenotypic characteristics. It can take as few as two to three consecutive seed to seed generations for some varieties to become irregular and non-uniform.

Small, Round and Icicle Types

The seed of modern, small, round and icicle types of radishes produced in the Pacific Northwest are planted in mid-spring when temperatures and day length are sufficient to produce rapid root growth and quality roots. The nursery can be planted anytime from mid-March till mid-April depending on location.

Roots should be pulled and selected at maturity. Once selected the roots may be planted immediately or stored in cool, high humidity conditions. Vernalisation requirements, necessary for flower initiation (bolting), are almost always met during the month of growth in mid-spring when they accumulate time of exposure to cool temperatures almost every night and sometimes during the day. In commercial seed production, the roots of these small-type radishes are often stored in a cooler at 35–42°F (2-5°C) for approximately 7 to 10 days. This insures that all of the radish roots in the population receive adequate vernalisation so they will bolt swiftly and uniformly upon replanting. This also contributes to the likelihood that all of the plants within the population will crosspollinate more evenly, as they will flower closer to the same time, and the seed will more uniformly.

Chinese, Black Spanish, and Daikon Types

Larger Chinese, Black Spanish, and Daikon radishes are true biennials that require more extensive vernalisation than the small types; however, vernalisation requirements vary by variety. Some may be produced by spring planting early enough (by late March) to acquire adequate vernalisation, while others are best seeded midsummer (late July) and overwintered. In most Pacific Northwest locations, the overwintered roots must be dug up and stored as they will freeze in the winter. Some varieties may prematurely produce a flower stalk before producing a marketable root when spring-planted. If root selection is required, it is preferable to plant these types in mid-summer (late July), dig and select the roots in the fall, store them through the coldest, wettest period of the Northwest fall and winter (November through January), and replant them as the day length increases in February. They will bolt readily as spring arrives and seed will be ready to harvest in mid- to latesummer, well before the first rains of September.

Spacing

Optimum spacing of a radish seed crop is a debated topic. Fully flowering radish plants, even those grown from petite, red types, can produce a plant that can reach 6 feet tall and 3 feet in circumference. Some growers plant at a wide spacing of 18 inches within the row and 4 feet between rows (row centers). This allows the plant to reach much of its seed bearing potential for maximum yields per plant in the field. The wider row spacing may be preferred as this will increase air flow through the crop, reducing disease pressures.

Alternately, some radish seed growers plant radishes as densely as 4 to 6″ spacing within in the row and with row centers that are 22 to 26″ apart. This tighter spacing can be achieved by transplanting roots at this density from the root nursery in the root-to-seed method or by directseeding very sparsely at this row spacing and then thinning the roots in the row to the appropriate spacing for the seed-to-seed method. With these higher density plantings, the yield per plant decreases but the overall yield per row foot may be comparable overall (or higher) as there are many more plants per row. Denser planting can prevent lodging and damage from wind in areas where this has the potential for being a severe condition at harvest time. The denser row spacing also has the benefit of fitting more roots within a given area, thereby increasing the population size.

Flowering and Pollination

Radish is a cross-pollinated species with perfect flowers. It is pollinated by a variety of pollinating insects attracted to the abundance of flowers over an extended blooming period. As with most other cultivated members of the Brassicaceae family, radish is self-incompatible (i.e., each plant is unable to pollinate its own flowers). Therefore, each pollination event comes from an outcross between two different radish plants of the same population. Initiation of flowering (bolting) in radish is not day-length sensitive, but most radishes do require some degree of vernalisation to induce flowering.

Isolation Requirements

As a cross-pollinated species radish can cross easily from one radish type to another. Wild radish, *Raphanus raphanistrum,* is a common weed in the Pacific Northwest and will cross with domesticated radishes. Isolation distances of at least one mile should be maintained from any other radish population if grown in an open landscape with few natural barriers. If barriers in the landscape are present, like forests, hills or large buildings, then it is possible to grow a radish population within 0.5 miles of another radish population with little chance of crossing. Isolation distances for seed production should also be increased if the nearest radish production field is of a completely different type. For example, if growing a round red type while another grower in the area is growing a Daikon type, then the isolation distance should be increased to two miles in an open landscape and one mile in an area with ample physical barriers.

Genetic Maintenance

Population Size

The extreme outcrossing (cross-pollination) of radish is ideal for maintaining adequate genetic mixing in a diverse varietal population. It is imperative to keep the level of inherent diversity in each variety intact when producing seed in order to avoid inbreeding depression. Seed should be collected from a minimum of 120 to 200 plants in any commercial radish seed crop.

Selection Criteria

The practice of genetic selection of any crop is related to the needs of the farmers in a particular area, the environmental pressures of the production region, cultural practices, and market demands. While breeders normally concentrate on traits for the farmers who will grow the crop as a vegetable, it is also important to remember that it is possible to select for traits important in seed production. Therefore, when developing varieties for organic systems selection criteria should include traits that are necessary for both the organic vegetable farmer and the organic seed producer. Selection should be done at several points in the life cycle, including at the seedling stage, the market or eating stage, and during flowering so as to maximise seed yield.

When approximately 80% of the roots in the root nurseries are of marketable size, the roots should be pulled and placed neatly on the beds for evaluation. Once the roots have been pulled for selection, it is best to keep them out of direct sun until replanted. Root selection on a cloudy, cool day is ideal. Selecting roots is best done with good knowledge of what the "norm" is for the phenotype (appearance) of each variety. Selection to a standard varietal type is the major objective in commercial production of established varieties. If a variety is adequately uniform at the outset of your seed production endeavor, then you can expect to discard ~20% of the roots at this stage in order to simply maintain varietal quality. If, instead, the variety is not uniform due to lack of prior selection, then expect to select out at least 50% of the roots as off-types. Selection for improvements is always best done at the market stage.

Radishes are commonly selected based on the following criteria in the course of seed production:

Seedling Vigor

Seedling vigor and early robust growth are important to organic growers as these traits affect weed competition, disease resistance, and overall plant health. These traits can be improved over several cycles of selection. Beyond selecting for the quickest or earliest germinating plants, early selection should include recognition of shape, size, color, and ability of the seedlings to grow under less than optimum conditions. This early selection should occur soon after emergence, and is best coupled with initial hoeing and thinning of the plot.

Leaf Size and Shape

While many growers do not pay attention to the leaves of radish, uniformity and quality of leaf in a radish variety tends to make more attractive bunches. Radish tops can range from 4 to 8" tall in both round and icicle types (even taller in winter types) so it is important that selection for height be done routinely. Radish tops can also vary between strap leaf and the lobed, cut leaf forms.

Root Size and Shape

Each radish variety has a distinct shape that is often somewhat different from other varieties and should be maintained. Select against off-type shapes and roots that are significantly smaller or larger. Shape can be quite fluid in radish, and offtypes will appear in each generation and must be selected against to maintain a good planting stock.

Root Color

Color variation in roots can be caused by previous outcrossing with a different colored variety, or because of inherent genetic variation. Regardless, roots should be selected for uniform color if the variety is to be maintained true to type. The ability to deepen the color is also possible, although this requires growing a large enough population to find a couple hundred roots of a darker hue to start a new vigorous population strain.

Harvesting and Cleaning

Seed Maturation and Harvest

Radish seed crops mature in approximately 150 days depending on variety, climate, and planting date. Radish seed is ready to harvest when about 60 to 70% of the seed pods turn from greenyellow to brown and lose their fleshy appearance, becoming papery thin and light. Plants are first cut above ground either mechanically with a drapelike swathe or by hand with a machete.

The plants are then placed in a windrow to cure in the field for 10 to 14 days. Unlike many other Brassicaceae crops radish pods do not shatter easily and only slight losses of seed occur during swathing and windrowing of the crop. The pods do not crack open readily and should be allowed to dry thoroughly to become brittle before attempting to thresh them. Combines or stationary threshers with rollers that can be set to effectively crack, but not crush, the pods are most efficient. Threshing can be done without machinery on a small scale by stomping on or driving over the pods.

Seed Cleaning

Once threshed, radish seeds may be separated from the pods, chaff, and debris by screening and winnowing the seed. Seed cleaning screens are numbered by number of 64ths of an inch size holes. Seed of most varieties may be separated by using a 9 or 10 size screen which allows the seed to pass through the screen while removing larger chaff. A 5 or 6 size screen may also be useful to retain the seed and allow smaller debris to be removed. If the chaff is adequately dried, much of it should be lighter than the seed and may also be removed by winnowing (separating by weight), by either dropping the material in front of a fan by hand or using a gravity table.

DISEASES

Several fungal and bacterial diseases are prevalent in radish seed production areas in the Pacific Northwest, particularly those that are favored by the wet conditions west of the Cascades Mountains. Due to the commercial demand for clean, pathogen-free seed, growers must pay particular attention to management of seedborne diseases. In organic production, the lack of approved pesticides and fungicides necessitates preventative management and non-chemical control measures. Routinely eliminating any exceptionally diseased plants in the field is important, not only for reducing incidence of infection, but for maintaining or improving the overall horizontal (or non-race specific) resistance to particular pathogens. Although it may take a number of cycles of selection to significantly increase the level of resistance in a variety, this slow, methodical approach can lead to improved resistance over time.

Critical seedborne and seed transmitted diseases which should be carefully monitored in radish seed crops grown in the Pacific Northwest include the fungal diseases Alternaria leaf spot *(Alternaria brassicae, A. brassicicola, and A.* raphani) and black leg (Leptosphaeria maculans, anamorph *Phoma lingam*), and the bacterial diseases black rot and bacterial leaf spot *(Xanthomonas campestris pv. campestris and X. campestris* pv. *raphani*). Starting with seed free from these pathogens is the first step in disease prevention. As all of these diseases are favored by moist conditions prevalent west of the Cascade Mountains in the Pacific Northwest, avoiding overhead irrigation aids in prevention or in minimising spread of these pathogens in the crops. Orienting rows and increasing row spacing to increase air flow through the canopy can also reduce disease development by reducing humidity and allowing the foliage to dry out quicker after rains or irrigation. Burying crop residue and controlling cruciferous volunteers and weed species are important as many foliar diseases persist and overwinter on plant tissue and debris. The pathogens listed above are not true soilborne pathogens, i.e., the pathogens will only persist in or on soils in host residues. Incorporating residues into the soil will promote more rapid decomposition of the residues than if the residues are left on the soil surface, reducing the duration of survival of the pathogens in the soil. Other diseases caused by true soilborne pathogens, i.e., Fusarium wilt and Verticilium wilt, necessitate even longer rotations. For this reason, a minimum 4-year rotation from all brassica crops is recommended. Additional disease-specific prevention or treatment measures are mentioned below, with accompanying disease descriptions.

Fungal Diseases

Alternaria leaf spot, gray leaf spot or black leaf spot (Alternaria brassicae, A. brassicicola, and A. raphani) – Seedborne

Alternaria leaf spot is caused by any of three *Alternaria* species, all of which cause similar symptoms on radish. The host is susceptible at all stage of growth, with cycles of infection favored by frequent and long periods of leaf wetness. On seedlings, stem lesions (dark spots) can result in seedling death. On leaves, lesions are tan, brown, or black spots of variable size; as they expand, lesions may be surrounded with a yellow halo, and lesions centers may show sooty black, concentric rings resulting from spore production. Symptoms usually show first on lower, older leaves. Lesions on flower stalks and pods are typically elongated compared to the rounder lesions that occur on leaves; infections on flower stalks also may allow bacterial soft rot to enter the stem, causing it to die prior to seed formation. Diseased pods may be distorted and dry prematurely, or be prone to shattering. Seeds in diseased pods may not form, or may become shriveled or blemished. Seedborne *Alternaria* can occur as mycelium growing under the seed coat, or as spores on the surface of the seed. Removal of infected leaves may slow progression of the disease. Burying crop debris is important as the pathogen overwinters on exposed plant tissue. Hot water treatment of seed can be an effective seed treatment.

Black leg (Leptosphaeria maculans, anamorph Phoma lingam) – Seedborne

The disease is caused by *Phoma lingam*. Symptoms are light-brown, oval to round spots that enlarge to become grayish-tan, frequently occurring near the bases of stems. Small black dots (fungal fruiting bodies called pycnidia or perithecia, depending on the stage of the fungus that is developing) form in the centers of spots (lesions); identifying these fruiting bodies is important for diagnosis of this disease, as this disease can readily be confused with ring spot, another pathogen of crucifer crops caused by *Mycosphaerella brassicola*. Black leg affects both foliar and crown tissues; wilted leaves are often a secondary symptom of severe stem and crown lesions. The fungus overwinters in crop debris or infected seed, with dissemination of spores occurring through splashing rain, irrigation, or wind. The fungus can survive for up to three years in the soil and can be seedborne. Disease development is favored by moist conditions, with greater severity in years with frequent rainfall in late spring and early summer. Avoid planting next to a field that has had crucifers in the past year; remove cruciferous weeds; incorporate crop debris; and avoid composts that may carry the pathogen as the disease may persist in the soil in decomposing host residues. Chose planting sites with good drainage. Hot water treatment can be an effective seed treatment.

Black root (Aphanomyces raphani)

The disease appears as brown to bluish-black discoloration in cracks or wounds on the side or top of roots. Root growth slows as symptoms progress; diseased roots become increasingly black and may cause girdling of the entire root. These black areas of discoloration extend inward in radial streaks. Choose planting sites with good drainage and avoid over-watering. Practice good field sanitation. Varieties vary in susceptibility, with 'White Icicle" being very susceptible. *A. raphani* is not seedborne, but resting spores of the pathogen can be moved in mud on tires or other equipment, and in trash or dirt in poorly cleaned seed.

Cercospora leaf spot (Cercospora cruciferarum and C. atrogrisea)

Cercospora leaf spot on radishes causes circular to angular grey or off-white lesions with darkened (brown to red-brown) margins on cotyledons, leaves, and petioles. The disease may eventually result in defoliation, particularly on younger seedlings. The pathogen survives in crop residues and can infect cruciferous weeds and volunteer plants. Tufts of spores are produced in the center of leaf spots and on overwintering crop residues. Rain or overhead irrigation is the most important mechanism of spore dispersal. Conditions for disease development are optimum in spring and late summer (temperatures 55Ý65°F (13–18°C) and infection is promoted by long periods of leaf wetness. Deep cultivation and removal of all cruciferous weeds/crops, along with crop rotations, may reduce incidence.

Fusarium yellows or Fusarium wilt of radish (Fusarium oxysporum f. sp. raphani = F. oxysporum f. sp. conglutinans Race 2) – Possibly seedborne

Fusarium yellows has recently been identified in a radish stock seed crop in the Columbia Basin of central Washington, causing complete crop failure. Although the disease had not previously been an identified problem in radish seed production in the Pacific Northwest, additional confirmed diagnoses of this disease causing severe losses to radish seed crops in the Willamette Valley of Oregon and the Columbia Basin of Washington, warrant careful monitoring and management because *F. oxysporum* has an extended period of viability in the soil (at least 10-15 years in some soils), the pathogen may be seedborne, and the disease can have devastating effects on a crop. *F. oxysporum* f. sp. *conglutinans* is host specific to crucifers, with several races identified. Race 1 infects a wide range of crucifers, including radish. Race 2 is able to infect radish and some other brassicas, but not *B. oleracea* crops like cabbage, cauliflower, and Brussels sprouts. Race 2 has also been designated as *F. oxysporum* f. sp. *raphani*. Symptoms appear as wilting and dieback of plants, often starting on one side of the plant of even one side of the leaf, accompanied by dark vascular discoloration. Root symptoms appear as a dark discoloration and rot. Severely infected roots readily become infested with cabbage maggot. Crop rotation is critical as the pathogen is soilborne.

Sclerotinia (Sclerotinia sclerotiorum) - Seed can be contaminated with dormant survival structures (sclerotia) of the pathogen

The disease causes necrotic, light to bleached lesions on leaves, petioles, and seed stalks. Continuous expansion of a lesion on the stem can girdle the stem, causing wilting and premature plant death of the plant above the lesion. A white cottony mold is commonly associated with the expanding lesions. Macroscopic, black survival structures called sclerotia are formed on and in diseased tissues. Sclerotia of *S. sclerotiorum* can survive in the soil for up to five years.

Germination of sclerotia leads to production of airborne spores and/or fungal strands that infect crown and foliar tissues. Symptoms typically are first observed on petioles of the flowers that have fallen and become trapped in the canopy (petioles serve as a source of nutrition for the fungus), and on leaves closest to the ground where the relative humidity is higher. Small sclerotia inadvertently harvested with seed, can contaminate seed lots and act as an inoculum source for future plantings. Rotate fields with non-hosts crops, such as grasses, for at least 4 years. Irrigate early in the day to provide sufficient time for plants to dry thoroughly in the afternoon sunshine. Remove infected plants prior to seed harvest.

White Rust (Albugo candida) – Seedborne

Symptoms first appear as light green spots that turn white and eventually develop into white, raised spore masses called pustules. Pustules, which may appear on upper or lower sides of leaves, contain white, wind-dispersed summer spores that can lead to additional infections. Some infections can become systemic, leading to the formation of galls, especially if the tissues of plants are very young. In flowering radish crops, *A. candida* causes bizarrely distorted seedpods, termed 'stagheads' (galls on infected seedpods). Stagheads significantly reduce seed yield and quality. Overwintering, resting spores of the pathogen (oospores) are produced in diseased tissues including the stagheads. Small pieces of stagheads can contaminate harvested seed and act as a source of inoculum. Oospores can also be seedborne. Disease progression is slowed by extreme hot or cold temperatures and by dry weather. In contrast, the disease increases quickly during mild periods with intermittent rains, heavy dew, and/or fog. Destroy crop residue and cultivate fields prior to planting to destroy all volunteer plants. Avoid planting in fields near other cruciferous crops. Hot water can be an effective seed treatment.

Bacterial Diseases

Black rot (Xanthomonas campestris pv. campestris, X. campestris pv. maculicola, X. campestris pv. raphani) – Seedborne

Radish is susceptible to all three pathovars of *X. campestris* that can infect crucifers. This can be a very serious disease of crucifers, including radish. Symptoms vary greatly depending on the age of the plant when infected, as well as the pathovar and the particular strain of the pathovar. *X. campestris* pv. *raphani* and *X. campestris* pv. *maculicola* attack the leaves and petioles, causing small tan to white spots with narrow, yellowish, water soaked zones on the leaves. The spots on the leaf petioles are black, sunken, and elongated. Severe infection results in defoliation and, in extremely severe cases, death may occur. *X. campestris* pv. *campestris* can cause angular, Vshaped yellow lesions that typically develop along the leaf margin, because the bacterium readily enters the leaf where the veins end at the leaf margin and form an opening called a hydathode. The veins in these yellow lesions may turn black. The lesions can dry up and turn brown. Severely infected leaves drop off the plant. Infection can turn systemic, i.e., the pathogen moves from the leaf into the petiole, the branches, and the main stem. If so, the vascular tissues turn black. Different races of the pathogen have been reported. In seed crops, infected plants may be symptomless, but systemic infection can lead to movement of the pathogen into the pods and seed.

The causal organism is carried over the winter in the crop residue and in infected seed. Once a plant is infected, further spread occurs via insects and splashing rain or irrigation water. During warm spring days, lesions are visible four to five days after infection. In cooler periods development is slower. The bacterium will grow between temperatures of 41–94°F (5–35°C), but is favored by temperatures between 80–86°F(27–30°C). Practice crop rotation for at least 2-3 years (longer for seed crops because infested stem tissues take a long time to be decomposed in the soil), and incorporate crop debris. Use seed that has been tested to be pathogen-free, or has very low levels of the pathogen. Seed treatment with hot water or chlorine can help reduce the amount of pathogen on the seed. Remove symptomatic transplants.

INSECT PESTS

Cabbage maggot (Delia radicum)

Adult cabbage maggots are small (¼ inch long) flies that lay their eggs on the soil surface near the radish crown. The larvae that hatch are ¼ inch maggots that tunnel into roots, causing wilting and stunting of plants. Maggots feed for up to four weeks, pupate in the soil, and emerge as adults. Pupae overwinter in the soil and begin emerging in late March. Multiple generations may occur in a season. Planting early will help avoid the first generation of flies. For extreme infestations, floating row covers will prevent flies from laying eggs. Cabbage maggots prefer moist, cool weather.

Cabbage seedpod weevil (Ceutorhynchus obstrictus)

The insect damages developing seed pods, reducing yield and providing entry for fungal pathogens. Adults are 1/6 inch long grayish weevils with a pronounced snout. Larvae are small white grubs that feed on developing seeds. One to two eggs are laid on the pods. These hatch within a week and the larvae tunnel into the pods. Larval development takes about 6 weeks. Adults feed on pollen and overwinter in leaf litter, ditches, etc. Populations may be reduced by colder winters with little snow cover.

REFERENCES

Koike, S.T., P. Gladders, and A.O. Paulus. 2007. *Vegetable Diseases*. A Color Handbook. Academic Press, San Diego, CA.

McIntosh, Charles. 2018.*"The Practical Gardener, and Modern Horticulturist."* Thomas Kelly: London. Page 288.

Sherf, Arden F. and Alan A. McNab. 2016. *Vegetable Diseases and Their Control*: Second Edition. Wiley and Sons. New York, NY.

Staff. 2006. *Pacific Northwest Plant Disease Management Handbook, Radish - White Rust*. Oregon State University Extension Service.

8

Spinach Seed Production

Spinach (*Spinacia oleracea*) is a member of the Chenopodiaceae or goosefoot family. It is derived from a leafy, winter annual that evolved in and near the Fertile Crescent of the Middle East. Winter annuals are plant species that germinate in the cool of the fall and grow vegetatively until the cold weather and short days of winter slow their growth. In spring, winter annuals grow steadily until a combination of environmental factors prompts the reproductive (bolting) phase of the life cycle.

Spinach bolting is initiated primarily by day length, and the ancestral forms of this crop bolt very early under less than 14 hours of light per day. This allowed the crop to mature seed before the onset of intense heat during summertime in the Middle East. The seed then lay dormant, having evolved to only germinate with the onset of cool, wet weather during fall. Modern forms of this crop have been selected to produce a lush and robust leafy vegetable that is more adaptive across environments and seasons, but some of the traits of spinach ancestors remain.

Many varieties of spinach still produce well as a fall-sown vegetable crop that is harvested in fall, winter, or spring. However, many spinach varieties have been developed to be springplanted and produce a bountiful crop before the summer day length causes bolting. Most modern European varieties do not initiate bolting until day length reaches close to 16 hours (the length of day in NW Washington near the summer solstice). While spinach seed can still be difficult to germinate when planted in warm soils, the seed is much easier to germinate than seed of the ancestral varieties.

GROWING SPINACH SEED

Climatic Requirements

Specific environmental conditions are required to produce good yields of high-germinating, largeseeded spinach without high incidences of seedborne pathogens. For this reason, there are few locations where spinach seed can be grown for commercial use. The two main spinach seed production areas worldwide are the Skagit Valley of Washington and a region of central Denmark. Both areas have cool, wet springs followed by dry, cool summers [temperatures usually not exceeding 75°F (24°C)] and relatively dry fall weather for harvest. Summer weather that exceeds 85°F (28°C), especially during pollination and early seed development, can dramatically lower germination rates, seed size, and yields. Skagit and Snohomish counties of Washington State have been major areas of production, with 2,000-3,000 acres of spinach seed crops grown per year at a volume equal to up to 50% of the US and up to 20% of the world production, according to the 2005 USDA "Spinach Seed Crop Profile" for Washington State. However, other areas in the Pacific Northwest, such as the north Olympic Peninsula, the San Juan Islands of Washington, and the islands of British Columbia have an even more moderate climate and are also prime locations for spinach seed production.

Figure 1. Spinacia oleracea

Soil and Fertility Requirements

Spinach grown for seed can be planted on a variety of soils, but the soils must be well drained to avoid root rot problems. Soil pH should be maintained above 6 as spinach is sensitive to acidic soils. In seed crops, the amount of available nitrogen should not be too high in order to avoid excessive vegetative growth before bolting, as this promotes lodging of the plants during seed set. Well maintained soils high in humus and microbial populations will supply adequate nutrients and water over the long seed production season. Spinach is somewhat tolerant of soil salinity and very tolerant of alkaline soils, although foliar fertiliser applications may be needed on alkaline soils to counteract the reduction in availability of micronutrients like manganese under high soil pH.

Planting

Most of the spinach that is grown in the Pacific Northwest for seed is planted in spring between March 15 and May 15. While spring-sown spinach can be planted as early as the ground can be worked, the plants will not put on any appreciable size until the onset of the longer days and warmer weather of late March or early April. Sometimes spinach seed crops are fall-planted and overwintered, but this is not commonly practiced due to increased disease pressure associated with fall-planting. Overwintered spinach develops a fuller canopy early in the spring which, combined with cool and damp conditions, can lead to increased incidence of some diseases, like Cladosporium leaf spot. Overwintering spinach also acts as a "green bridge" providing a vegetative host allowing many diseases to survive through the winter.

The temperature range for germination of spinach seed is 45-75°F (7-23°C) with an optimum of 70°F (21°C). Hot weather can suppress germination. Daily irrigation during hot periods can help cool the soil and facilitate germination. Shade cloth or Remay can also be used to cool the surface of the soil during hot periods.

Spinach seed crops are commonly grown with overhead irrigation; however, drip irrigation may reduce disease pressure. In the Skagit Valley of Washington, most spinach seed producers do not irrigate, relying instead on rainfall. In the Willamette Valley of Oregon and other regions of the Northwest irrigation is required.

Spacing

The planting density for spinach seed production requires much wider spacing than for vegetable production. In general, spinach plants should be spaced 8 to 12 inches apart within rows. If the population is to be selected for genetic maintenance or improvement then the initial planting density is anywhere from six to eight plants per foot to allow for multiple selection events from time of emergence to flower initiation. Standard row centers are normally 22 to 26 inches, but in raised beds, spacing between rows can be dropped to 12 to 14 inches. Wider plant spacing increases air flow through the crop, reducing disease pressure.

Cultivation

Spinach seedlings grow slowly and do not compete well with weeds. Starting with a stale seed bed is advisable, as well as avoiding fields with high weed pressure. Pre-emergence flame weeding may be effective, but emerging seedlings are easily damaged. Early season weed control is critical for optimum establishment. If plants become leggy due to early competition they will be difficult to evaluate in the selection process, produce lower yields, and be more prone to lodging when full size.

FLOWERING AND POLLINATION

Spinach is a wind-pollinated crop with extremely fine pollen. It is a dioecious species: approximately half of the plants in a population have only male flowers, while the other half have only female flowers.

Occasionally, there may be monoecious plants in a spinach population that express both male and female flowers on the same plant, but these are less common. Male spinach plants exhibit two basic plant forms. The first males to flower are quite short in stature (often only 4 to 6 inches tall at full development) with suppressed leaf development on all upper nodes, but prodigious in staminate flowers at all nodes. These "extreme males", as they are often called, only flower for a short period of time, but produce large amounts of pollen ensuring that plenty of pollen is available for early flowering females. The second type of male plant is known as a "vegetative male", and is more typical in morphology with both male flowers and leaves at all nodes. Vegetative males initiate flowering a week to 10 days after the extreme males. These plants flower over a longer duration, ensuring ample pollen for the female plants throughout the period of pollen receptivity. In the absence of pollen, female plants eventually "revert" and begin producing male (pollenproducing) flowers.

Flowering is initiated primarily due to day length. Heat can play a role by speeding the metabolic rate of the spinach plant, accelerating the flowering process once flowering has been initiated.

Hybrid Seed Production

Most commercial spinach varieties including an increasing number of organically available varieties are hybrids. The fact that spinach is a dioecious species is an advantage in hybrid spinach seed production. The female plants are used to develop all female flowering parental lines. This is possible because female spinach plants, if isolated from spinach pollen during flowering will eventually "sex reverse," subsequently producing male flowers in addition to their female flowers in order to achieve pollination. The subsequent generation will inherit the female genes. This sex reversal enables breeders to increase seed of a female line. During the development of female lines for hybrids the breeder must perform meticulous selection on the female plants insuring that the lines uniformly produce only female flowers for at least five to six weeks before sex reversing. Then, when used as female parents in hybrid seed production, these lines will have adequate time to receive all of their pollen from the male parental line of the hybrid in the field.

Dioecious, open-pollinated populations may also be used as "females" in hybrid seed production, however all male plants must be removed from the population prior to flowering to ensure pollen is received from the intended male parental line. Monitoring and removing male plants is also required when using an all female line as there is always a chance that although predominately female, a male plant may remain in the genetics of the population.

Traditionally, the male parents for hybrid spinach seed production have been populations of spinach varieties with the normal complement of approximately 50% male plants and 50% female plants. The male parents are usually more genetically elastic than their female parental counterparts, in part to maintain their vigor and ability to produce ample pollen, and are therefore referred to as "open-pollinated parents" or "OPs" among spinach seed producers. The male plants in these OPs tend to have the two distinct forms, "extreme males" and "vegetative males" (described above).

As 50% of the plants in the male parent produce ample pollen producers it is not an issue that the other 50% of the plants are female and do not contribute directly to the production of the hybrid. The male rows are usually destroyed several weeks before the hybrid seed is harvested from the female, seed producing rows, to avoid any chance of seed mixing at harvest. In recent years there has been a trend to develop male parental lines that are monoecious, having both male and female flowers on the same plant. These monoecious male parents are derived through inbreeding and tend to be more uniform in their reproductive characteristics than the traditional dioecious males. Also the elimination of extreme male plants from these monoecious males means there is later flowering of these male parents. This trait is passed on resulting in later bolting in the hybrids produced by these lines. The main reservation with the use of monoecious males at this time is that many of them do not produce as much pollen as the traditional, dioecious males.

ISOLATION REQUIREMENTS

Spinach readily cross pollinates, so an isolation distance of at least 1-2 miles between adjacent spinach seed crops is recommended. Distances may be adjusted according to direction of prevailing winds, similarity of neighboring spinach crops (for example savoy vs. flat leaf cultivars) and the presence of wind breaks or barriers. Check with local state extension offices about the use of regional pinning maps to determine the location of spinach seed crops. If pinning maps are used in the area, growers are requested to post the location of the spinach seed crop after checking for appropriate isolation distances from other spinach seed fields in the area.

GENETIC MAINTENANCE

Population Size

Seed should be harvested from a minimum of 120 plants during selection in order to maintain seedling vigor and avoid inbreeding depression in spinach breeding programs. A larger population, 200 plants or more, is recommended in order to maintain good genetic diversity. The initial population size should reflect the intensity of selection activities.

Selection Criteria

The practice of genetic selection of any crop is related to the needs of the farmers in a particular area, environmental pressures, cultural practices, and market demands. When producing seed from stock seed (seed that has already been genetically refined) or under contract from a seed company, minimal roguing or selection may be required. Seed company guidelines should be communicated and followed regarding timing and extent of roguing activities. However, more intensive selection is an effective tool for improving cultivars over time and may be particularly useful in adapting a cultivar to local, organic growing conditions. While breeders normally concentrate on traits for the farmers who will grow the crop as a vegetable, it is also possible to select for traits important in seed production, a crucial component in adapting cultivars to organic systems. Selection should be done at several points in the life cycle. Planting six to eight spinach plants per foot in the initial stand before selection establishes a population large enough to select, or rogue, intensely and still maximise crop yields.

Spinach populations are commonly selected based on the following criteria:

Seedling Vigor

Seedling vigor and early robust growth are important traits in all production systems, and can be improved over several cycles of selection.

This includes not only selecting for the quickest germinating plants, but also recognising based on shape, size, and color, the ability of seedlings to grow under less than optimum conditions. This is done soon after emergence and is best coupled with an initial hoeing of the field.

Leaf Shape

There are various stages at which leaf size can be selected depending on the leaf shape desired by the vegetable market served by the spinach cultivar. As an example, much of the modern "baby leaf" is selected for rounded leaves. Alternately, some prefer a lobed or arrow-shaped spinach leaf for shape variation. Leaf textures are characterised as savoy (crinkle-leaf), semi-savoy or flat, and are selected along with leaf shape. Leaf texture varies with the stage of growth.

Leaf Color

Color variation exists in spinach and may be selected. Darker leaf color is commonly preferred. Darker leaf color is also attributed to higher nutritional value. Select at various growth stages for color.

Plant Stature

The ability of the plant to hold its leaves in an upright position is very important in vegetable production in general, especially for harvest of leaves cut for salad mix. Upright foliage also lessens the amount of soil trapped in the underside of leaves and can reduce the amount of fungal and bacterial pathogens splashed onto leaves from rain or irrigation.

Disease Resistance

Diseases should be monitored and identified accurately for proper field management and for selection purposes. Routinely selecting for nondiseased plants can help develop partial or horizontal disease resistance. This requires familiarity with the symptoms of various diseases, and selecting the plants that exhibit the least severe symptoms (specific diseases of spinach are covered in the *Disease* section below). Horizontal resistance is a durable form of resistance that combines multiple disease resistance genes into the population. Although this type of resistance may still incur some level of disease in the crop, the concept is to reduce the incidence of disease to a tolerable level.

Early Flowering

The earliest flowering males (usually "extreme males") are also the earliest flowering plants in spinach populations. If all "extreme males" are selected out of the population along with the earliest flowering females then it is possible to shift the spinach variety to a longer-standing, more bolt-resistant, variety.

Cold Tolerance

There is usually minimal cold weather pressure on overwintering spinach populations in NW Washington. However, there have been several very cold winter episodes in the past decade that have resulted in considerable damage to overwintered spinach crops. Above 18°F (–8°C), damage is usually minimal. Below 15°F (–9 to –10°C), damage can be severe and crop loss can occur. In the temperature range of 15–18°F (–8 to –10°C) damage is variable, allowing for selection of the least damaged individuals. Hardier varieties that are more adapted to Pacific Northwest winters can be developed through several cycles of such selection.

SEED MATURATION AND HARVEST

One of the most difficult steps in successfully growing a seed crop is judging the maturity of the seed. If a crop is harvested before the seed is fully formed, the seed may be immature, have low germination, and lack seedling vigor. If the crop is harvested too late, there is the risk of losing seed to shattering, predation, or inclement weather. Prolonged exposure to wet weather can also promote bacterial or fungal diseases that may degrade the quality of the seed.

Spinach seed, as with all members of the Chenopodiaceae, is formed and matures in an indeterminate growth pattern, beginning on older, lower branches and continuing up the flower stalk through the season. Due to this sequential maturation, only a portion of the seed continuously being set will reach maturity by the end of the season. In most cases only about 75% of the seed on any given plant will reach maturity by harvest. One method of gauging the maturity of a spinach seed crop is to make a visual assessment of the percentage of seed on most plants that has turned a tan-brown color, typical of mature spinach seed, and then harvesting the crop when 60 to 80% of the seed has become this color. However, this is an unreliable method of judging maturity due to the considerable effects of genetic and environmental variation.

Environmentally, the seed may prematurely lighten in color in the presence of the Stemphylium/Cladosporium leaf spot disease complex. Genetically, some spinach cultivars have much greener seed, even at full seed maturity.

The most important trait to monitor in gauging the maturation of spinach seed is the relative maturity of the seed endosperm. The endosperm and embryo, which mature concurrently, must be fully developed to produce vigorous, vital seed. The endosperm, which is primarily comprised of starch, is easily monitored by cracking open or squeezing the seed to visually inspect the stage of endosperm development. Appearance of the endosperm will transition from translucent or milky early in development to what is described as "flinty" with a grayish, waxy appearance at the midpoint of development. At these stages, the endosperm can be squeezed out of the developing seed for inspection. When seeds in the middle of the stalk are at the "flinty" stage, irrigation may be stopped to speed the maturation and drying process. This can be important if there is risk of maturation extending into the wet, diseaseconducive conditions of fall.

At a mature stage of development, the endosperm turns a starchy, white color which can be observed by cracking open the seed. When the endosperm turns to this true, solid white color and is firm, the seed is mature and ready to harvest. When the majority of the plants in the field have at least 75% of the seeds at this advanced, starchy white stage, it is time to cut the plants near the base of the stems to stack into windrows. This should be done preferably during a warm, dry period. The windrowed stalks will be ready to thresh in 4-10 days, depending on the weather. Rotating stalks in the windrows facilitates uniform drying of the seed. Cutting or windrowing the crop in wet weather should be avoided. If a crop is ready to cut and an extended wet weather pattern is forecast, cut the crop and bring it into an airy, dry shelter to cure. Once dried, seed stalks are usually combined or threshed and then cleaned by screening and fanning.

DISEASES OF SPINACH SEED CROPS

Several fungal diseases are prevalent in spinach seed production in the Pacific Northwest, particularly those that are favored by the wet conditions of the region. Due to the commercial demand for clean, disease-free seed, growers must pay particular attention to management of seedborne diseases. In organic production, the restriction of available pesticides and fungicides necessitates preventative management and nonchemical control measures. Spinach leaf spot diseases, including anthracnose, Cladosporium and Stemphylium leaf spot, and downy mildew are key fungal diseases which thrive in cool to warm, wet conditions, may be seedborne and are prevalent in the Pacific Northwest. They may occur individually or as a disease complex. Fusarium wilt, another fungal disease, has the greatest impact on spinach production in the spinach growing regions of the Pacific Northwest. Although favored by warmer temperatures, the pathogen thrives in moist conditions, can be seedborne, and is long lived in the soil. For this reason, growers must practice long rotations (5-17 years) from spinach and related host plants. Preliminary research suggests that *Fusarium oxysporum* f. sp. *spinaciae,* the causal agent of spinach Fusarium wilt, is favored by lower pH (acidic) soils, so the addition of lime to soils prior to planting may aid in suppression of Fusarium wilt.

Cultural controls, particularly important in organic production, play a large role in disease prevention. Cultural recommendations to prevent an array of spinach diseases include managing irrigation to minimise splashing and the durations of leaf wetness; destruction of field inoculum by incorporating infested residues into the ground; crop rotation; row orientation, staking, and increased row spacing to increase air flow in the field and reduce relative humidity; planting pathogen-free seed; and the use of diseaseresistant cultivars. Spring planting rather than fall planting is also commonly recommended in the Pacific Northwest to reduce disease pressure. Fallplanted spinach has a fuller canopy in the early spring when moist, cool conditions and lack of air flow can increase disease pressure. Additional disease-specific recommendations along with disease descriptions are listed below.

Hot water seed treatment has been shown to reduce the infection of Cladosporium variabile, *Stemphylium botryosum* and *Verticillium dahliae* in spinach seed. However, the specific temperature and duration of hot water treatment must be controlled carefully to avoid damaging the seed and reducing germination. Research at WSU by Dr. Lindsey du Toit demonstrated eradication of *C. variabile* from spinach seed treated with hot water at 40°C for only 10 minutes, and *V. dahliae* at 55°C for 30 minutes or 60°C for 10 minutes without reducing germination. *S. botryosum* was eradicated from a lightly infected seed lot treated at 55–60°C for 10 minutes without reducing germination, but could not be eradicated from a heavily infected seed lot, even by treatment at 60°C for 40 minutes. Hot water treatment can affect longevity and germination rates in seed, so the procedure should be used with caution. Experiment with small batches at a time.

Fungal Pathogens

Downy mildew (Peronospora farinosa f. sp. spinaciae) – Seedborne

Downy mildew is a wide-spread disease in spinach in many regions of the U.S. The disease is favored by cool, wet conditions, and can be devastating in spinach seed or vegetable production. The disease is characterised by slightly yellow, irregular, chlorotic lesions on leaves that may progress into expanded necrotic patches. There is evidence that downy mildew can be seed-transmitted, but the rate of seed transmission and impact of seed inoculum is unclear. Fungal spores spread readily by wind and splashing water. Heavily infected leaves may become curled and distorted. Incorporation of infested crop residue into the ground is recommended as the pathogen overwinters on spinach debris and weeds, but is not a soilborne pathogen, i.e., the pathogen does not survive once the crop residues have been decomposed by the soil microflora/fauna.

Anthracnose (caused by Colletotrichum dematium f. sp. spinaciae) – Seedborne

Anthracnose is characterised by small, circular, water-soaked lesions on young and old leaves which can enlarge and become chlorotic or necrotic. Black fruiting bodies called acervuli form in lesions and distinguish the disease from other leaf spot diseases. The fungus thrives in wet, cool conditions and is dispersed by splashing water and on infected seed. Incorporation of crop residues is recommended as the disease overwinters on infected spinach debris and seed.

Stemphylium leaf spot (Stemphylium botryosum)–Seedborne

The disease is characterised by distinct spots on leaves that begin as a grey-green color, turn tan, and then expand rapidly and become dry and papery. The fungus thrives in moist, warm conditions and is more aggressive on spinach in the presence of pollen. Spores are produced in lesions on leaves in moist conditions. The fungus is dispersed by wind and on infected seed. Incorporation of infested seed crop residues is recommended as the pathogen overwinters on infested residues left on the soil surface, and on infected seed. Orienting rows and increasing spacing to improve airflow are recommended, in combination with planting pathogen-free seed.

Fusarium wilt (Fusarium osysporum f. sp. spinaciae) – Seedborne

Fusarium wilt is characterised by a general wilting of plants, a flaccid, off-green color, and blackened roots externally and within the vascular tissue. The disease can occur at all stages of growth. It is favored by warm conditions when the transpirational demand of the plants is greater but the vascular tissue is compromised because it is blocked by the fungus. The pathogen can survive in the soil in the absence of a host for many years. Crop rotations of five to seventeen years from spinach, beets, and chard are recommended. The duration of the rotation depends on cultivar susceptibility and the amount of field inoculum. Preliminary research suggests that *F. oxysporum* f. sp. *spinaciae* is suppressed in high pH soils, so the practice of liming soils prior to planting may be beneficial for control of this disease.

Damping off and root rots

Damping-off is a prevalent soilborne disease in spinach caused by various pathogens, primarily species of *Pythium, Rhizoctonia* and *Fusarium*. These pathogens are not seedborne, except for some of the *Fusarium* spp. Damping-off symptoms include lesions, water-soaking, and/or girdling of the roots and crowns that result in wilting and death. The disease is favored by wet conditions and only affects seedlings and young plants. Avoiding overwatering is important in prevention of the disease, particularly for *Pythium* spp. (water molds). Although disease resistant cultivars are not commercially available, there is evidence of some resistance in cultivars suggesting the potential for development of horizontally resistant cultivars.

Viral Pathogens

Cucumber Mosaic Virus (CMV) – Seedborne

A virus, CMV causes a disease of spinach also referred to as spinach blight. Symptoms range from light to severe chlorosis and mosaic patterns of the leaves, and can include narrowed crown leaves that become curled, wrinkled, and roll inward; yellowing of leaves that may die; and stunted plants. The pathogen is aphid-vectored and overwinters in perennial weeds and vegetables (especially cucurbits). Some strains are seedborne and seed-transmitted. Managing aphids and avoiding planting near cucurbit crops is recommended. Roguing symptomatic plants in the field can help prevent spread of the disease.

Beet Curly Top Virus (BCTV)

BCTV is a virus that affects many plant species and is vectored by leaf hoppers. The disease is characterised by thickened, rolled leaves that may be yellow, brittle, and stunted. Affected plants appear compact and stunted, and may die. Control of leaf hoppers and not planting near other Chenopodiaceous crops is recommended.

Beet Western Yellows Virus (BWYV)

BWYV is a virus that has a very broad host range, is particularly prevalent on brassica weeds, and is vectored by aphids. Symptoms of infection include interveinal and marginal yellowing, appearing first on older leaves; red-brown spots between veins that have a bronze cast; thick, leathery, brittle leaves; and poor root growth. The disease is readily confused with symptoms of nutrient deficiencies (e.g., nitrogen, iron, or magnesium). Managing aphid populations is recommended.

REFERENCES

Corell, J. C., Morelock, T. E., Black, M. C., Koike, S. T., Brandenberger, L. P., and Bainello, F. J. 2014. "Economically important diseases of spinach". *Plant Dis.* 78:653-660.

Doijode, S. D. 2021. *Seed Storage of Horticultural Crops*. Hawthorn Press. New York, NY.

du Toit, L. J. and Hernandez-Perez, P. 2015. Efficacy of hot water and chlorine for eradication *of Cladosporium variabile, Stemphylium botryosum* and *Verticillium dahliae* from spinach seed. Plant Dis. 89:1305.

Foss, C. R. and Jones, L. J. 2015. Crop Profile for Spinach Seed in Washington. USDA National IPM Center. http://www.ipmcenters.org/cropprofiles/docs/ was pinachseed.html.

Hernandez-Perez, P. and du Toit, L. J. 2006. Seedborne *Cladosporium variabile* and *Stemphylium botryosum* in spinach. Plant Dis. 90:137.

Jarmin, M. 2019. *Spinach Seed Production in the Pacific Northwest*. Washington State University Extension Publication: PNW 267.

Koike, S.T., Gladders, P., and Paulus, A.O. 2007. *Spinacia oleracea*. Spinach. Pages 368-381, in: Vegetable Diseases. A Color Handbook. Academic Press, San Diego, CA.

Morelock, T. E. and Corell, J. C. 2006. *History of Spinach Production in the United States*. Proceedings from the 2006 International Spinach Conference, 13-14 July 2006, La Conner, WA.

Yang, Y., Soo Kim, K., and Anderson, E. J. 2017. Seed transmission of *cucumber mosaic virus* in Spinach. Phytopathology 87:924-931.

9

Seed Production of Tree Plants

The use of well-filled viable seed of good inherent quality provides a sound basis for the raising of vigorous and healthy forests capable of producing wood of good quality. The efforts of forest geneticists and tree breeders to introduce superior provenances and improved cultivars into forest practice depend on the organisation of an adequate supply of seed for which special seed producing units are required. Seed of the best provenances is produced in selected parts of natural forests or selected plantations which have been converted into seed stands or seed production areas; while seed of new improved cultivars is produced in seed orchards which consist of grafted plants or plants of seedling origin, derived from selected parent trees.

Besides being of a superior provenance or cultivar, the seed which reaches the forester for sowing in the nursery or in the forest must be clean, viable and healthy if wastefully low yields of plants or crop failures are to be avoided. In more precise terms, the seed must be free from impurities, possess high germinative capacity and be free from seed-borne pests and diseases. Thus, if the productivity of the forest is to be raised through the use of the products of selection and breeding, the supply of seed and plants must be so organised and regulated that it is possible to obtain seed and plants that are both true-to-name and satisfy certain minimum requirements of quality.

SEED PRODUCTION IN FOREST TREES

Seed production areas and seed orchards exist to produce the greatest possible yield of well-filled viable seed. Their successful management depends on a thorough knowledge of the genetics and physiology of flowering and seed production, and it is therefore necessary to summarise the present state of knowledge on these subjects.

Flower Bud Initiation

In many temperate tree species flower primordia are formed early in the growing season preceding the spring in which the flowers appear. The critical period appears to lie between early May and the end of July. A certain minimum degree of heat is apparently necessary for flower bud initiation. Trees grown in sunny positions come into flower earlier in life and flower more regularly and prolifically than those grown in a close stand or in shade. A reduction in water supply in summer is frequently associated with flower bud formation but lack of moisture evidently can become a limiting factor and cases are known where irrigation resulted in an increased seed crop. Sarvas has provided ample evidence for *Pinus sylvestris* that flowering and seed production are better on the more fertile sites. Although similar detailed studies of the relationship between flowering, seed production and high soil nutrient status are lacking for other species, it is now generally accepted that a relatively high soil nutrient status is essential for the regular initiation of flower buds and that vegetative vigor *per se* does not necessarily preclude flower bud initiation.

There is no doubt that on many forest sites the application of fertilisers will stimulate flowering and seed production but the results obtained so far have been very variable. Responses in flowering and fruiting have been obtained from the application of nitrogen, phosphorus and potassium applied singly; from combinations of nitrogen and phosphorus, and phosphorus and potassium; and from complete NPK fertilisers. In at cases where the fertiliser was applied in combination with crown release by thinnings, the combined treat meets produced better results than were obtained with fertiliser alone, and the same obtains where fertilises has been applied in combination with irrigation.

Little is known of the role of plant growth substances and the internal changes in forest trees associated with flowering, but the success of Japanese workers in using gibberellic acid to induce flowering in a number of coniferous tree species encourages further research into the role of plant growth substances in flower initiation.

Pollination

The importance of adequate pollination for the development of viable seed by forest trees has been well demonstrated for several species. Cross pollination is the general but by no means universal rule in many tree species of economic importance, and a considerable number are more or less self-sterile.

It now appears that the dispersal of pollen in wind-pollinated species is not so dependent on the weather prevailing during the period of flowering as was once thought; the reason appears to lie in the great speed and quantity of pollen dispersal as soon as conditions are right. Extensive observations on the flowering of *Betula verrucosa, B. pubescens, Picea abies* and *Pinus sylvestris* in Finland led Sarvas to conclude that the start of flowering, the duration, and the occurrence of the peak of flowering are all closely dependent on temperature. Strong winds may accelerate pollen release and rain may interrupt it, though rain seems to have less effect on pollination and seed set than has often been assumed. In the case of insect-pollinated species, such as *Acer pseudoplatanus* and *Tilia*, good flying weather for insects is required for pollination, that is bright, dry and fairly calm weather.

There are many records of small amounts of airborne tree pollen traveling great distances from the source (" foreign pollen ") but several workers have shown that the bulk of airborne pollen falls fairly close to its source (" local pollen "). The results suggest that pollination is mainly effected by trees in the immediate vicinity of the seed tree. However, the behavior of pollen produced by stands of trees must also be considered as well as factors affecting the distance of dispersal including the size and shape of the stand, and the quantity and nature of the pollen produced. Pollination conditions are best in extensive stands containing trees with well-developed crowns.

Seed Formation

Pollination is normally followed by fertilisation. The process of pollen formation, pollen shedding and sub sequent germination on the tip of the nucellus in *Larix* and *Pseudotsuga* is described in detail by Barner and Christianson. The lapse of time between pollination and fertilisation is usually quite short, but in *Pinus* the interval is about one year.

The comprehensive study made by Sarvas can again be referred to for a good account of the normal course of fertilisation in *Pinus sylvestris*. The pollen grains are normally very viable, they develop sperm nuclei with great regularity and these are carried with great accuracy to the eggs by the pollen. The eggs are generally fertile and there appear to be no genetic barriers that significantly prevent fertilisation. However, the competition between developing embryos is often decided by genetic factors as, for example, when the union of semi-lethal genes to form homozygotes at fertilisation leads to the defeat of the embryo during the competition in the embryo cavity.

Nonpollinated ovules gradually collapse, mostly during the first growing season and in *Pinus* poorly-pollinated cones soon drop. High embryo mortality in *Pinus sylvestris* is attributed by Sarvas to self-fertilisation, and Ehrenberg *et al.* also showed that self-fertilisation often results in poor germination of full seeds and inferior progenies.

Size of Seed Crop

Trees that bear heavily one year and then sparsely or not at all for several years are said to show periodicity in seed-bearing. In many species the seed crop of one year affects the seed crop of the following year, both through the reduction of potential foliage-producing or vegetative buds and the reduction of carbohydrate resources. The flowering and seed production of young trees is commonly sparse and sporadic but increases rapidly with increasing age and size. Within a forest stand it is the larger, dominant trees that bear the most seed but there are also inherent differences between trees in flowering, fruiting and the production of viable seed that make necessary careful selection for fertility and fruitfulness.

Good flower years occur much more frequently than good seed years. There may be several reasons why a spate of flowers often fails to produce a spate of seed, but deficient pollination, the effects of self-fertilisation, the activities of seed-destroying and cone and fruit-damaging insects, animals and birds and unfavorable weather are usually to blame for this. Periodicity in flowering and seed production depends to a large extent on external factors and protection from climatic dangers and the animals, birds, insects and fungi that damage or destroy flowers, fruit and seed should do much to mitigate, but not prevent irregular seed production. For many species the gap between good seed years can be bridged by the use of refrigerated seed storage.

Quality of Seed Crop

The spells of warm, dry sunny weather which appear to favor the initiation of flower buds and the accumulation of carbohydrate resources also favor the maturing of the fruit and seed. Considerable variation in seed quality and time of maturation has been found between individual trees and stands of the same species.

In northern latitudes a greater part of the cone or fruit crop is borne on the south than on the north side of the crowns of seed trees. This distribution affects the methods used to assess the size of seed crops. In many species the best cones, fruits and seeds are generally found on the middle and upper parts of the crowns. There is also a variation in seed quality within the cones of many coniferous species, the middle portion being the most productive of well-fired viable seed.

Two general recommendations for the improvement of seed supplies can be made.

1. Full advantage should be taken of good seed years in all species, because the highest proportion of well-filled viable seed is produced in such years.
2. Much can be gained by improving seed collection methods both from standing trees and from the ground as in *Fagus, Quercus* and *Tectona.*

TREATMENT OF SEED STANDS

Choice of Seed Stands

The objects of forming seed stands or seed production areas are to:

1. produce seed of improved inherent quality by selecting and favoring seed trees which are vigorous, straight-stemmed and healthy and produce wood of good quality;
2. concentrate seed collection into a few specially treated parts of the forest, thus making seed collection easier to organise and control;
3. improve the germinative energy and germinative capacity of the seed collected.

All three objects can be achieved by one simple treatment, that is, the careful selection of seed trees and the complete isolation of the crowns of these seed trees by thinnings.

A necessary preliminary to the formation of seed stands is a survey of the region to locate suitable natural stands or plantations. A seed source is a crop of fast-growing and healthy trees with good growth habit suitable for seed production. Plus seed sources are suitable for regular and intensive seed collection; normal seed sources are suitable provided careful selection is made of the seed trees; minus stands are unsuitable for seed collection. A seed source becomes a seed stand or seed production area when all defective trees have been removed, leaving the best trees isolated to develop big crowns and so produce more seed. The ground and the seed trees may also be treated to increase the seed crops and make collection easier.

Seed stands are usually formed in natural forests or plantations that are sufficiently old or sufficiently well-developed to provide reasonable assurance that they are well adapted to the site and will continue to show rapid and healthy growth. A suitable lower limit is a top height of 11 to 12 meters (36 to 39 feet) and a good upper limit is half the rotation age. Clearly it is desirable to convert relatively young stands because the response of younger trees to thinnings and fertilisers is greatest and there will be opportunity as the trees develop in later years to make further selection among the seed trees. Nevertheless, many outstanding or plus stands approaching or at the age of maturity for timber purposes can form valuable seed stands. The area of seed stand required can be fixed by observing the average seed production of well-grown trees and setting this against the future planting or regeneration program.

Isolation from Inferior Sources of Pollen

An additional criterion in choosing seed stands is the absence of neighboring natural stands or plantations of the same or closely related species which are of poor or minus quality. Such isolation from inferior sources of pollen is very important. Dispersal distance is affected by very many factors and it is easy to oversimplify a complex situation. Nevertheless, when fixing the size of isolation strips around the seed stand and isolation distances from large sources of undesired foreign pollen, it should be remembered that the degree of contamination by foreign pollen is roughly proportional to the size of its source, and that contamination is least when the size of the seed stand and the local pollen cloud are large in relation to the cloud of foreign pollen. These considerations are the basis for the recommendation that seed stands should be at least 3 and preferably 5 hectares (7.5 to 12 acres) in extent.

First Step in Forming Seed Trees

The first step in forming a seed stand is to select and mark the seed trees. These should be fast-grown healthy dominants capable of producing wood of good quality. The stems should be persistent and straight, and also free from defects such as fluting, spiral grain and epicormic shoots. The branches should be smart in relation to the stem at the point of origin, the branch angle should be flattened to moderately ascending, and good natural pruning is also a desirable feature. The crowns should be compact and well-provided with foliage and the trees should show signs of having borne seed in the past. Where wood characters such as specific gravity or tracheid and fiber dimensions can be clearly specified, it may also be possible to select seed trees for these characters.

The number and relative superiority of the seed trees marked have an important effect on the genetic improvement of the progeny raised from the seed collected in the seed stand. More precisely, the genetic gain or change in the average genotype of the population produced by the selection of seed trees depends on the heritability of the characters for which selection is made, the variability of these characters and the proportion of the population selected. The genetic gain from seed stands is greater when the proportion selected is low, but the number of seed trees per hectare must be sufficient to ensure adequate cross-pollination between the seed trees, thus avoiding the effects of self-pollination or inbreeding.

Thinning Seed Stands

The second stage in forming a seed stand is the removal of the phenotypically inferior trees to free completely the crowns of the seed trees. Sometimes seed stands are formed from groups or avenues and such trees often have long, well-developed crowns. In most cases, however, the seed stands will previously have been managed for wood production so that the crowns of the seed trees are relatively small. This is particularly the case in the light-demanding species of *Pinus* and *Larix* and also holds for *Tectona grandis, Eucalyptus* species and *Casuarina*. The frequency and intensity of thinnings to achieve full crown release must depend on local conditions, and particularly the danger of windthrow, but the sooner the seed trees are isolated the better for future seed production. The first thinnings should be in the nature of heavy crown thinnings to free the crowns of the seed trees, and to remove large, rough dominants from the seed stand and isolation strip. Thereafter the remaining trees are removed as soon as convenient or safe.

Fertiliser Treatments

The next step is to consider what other treatments should be applied to increase flowering and seed production still further and to make their occurrence more regular. Root pruning and stem girdling or strangulation are likely to be more harmful then helpful, but the application of fertilisers stands out as a beneficial and worthwhile treatment, especially where soil fertility is relatively low.

The quantities and relative proportions of the various major elements which are necessary to stimulate flowering and seed production obviously vary with the nature of the crop, especially the age and size of the trees and the site conditions. It is, however, clear that the fertiliser should correct existing deficiencies and supplement the major nutrients and for this a general prescription such as 2N: 1P: 2K at the rate per hectare of 112 kilograms of nitrogen, 56 kilograms of phosphorus and 112 kilograms of potassium (or about 100 pounds of nitrogen, 50 pounds of phosphorus and 100 pounds of potassium per acre) would appear to be a good starting point for fertiliser treatments. Usually the fertiliser is applied over the whole area but where the seed trees are scattered it is preferable to apply it below each seed tree to an area of ground one-and-a-half times the diameter of the crown.

In general the best time to apply compound fertilisers is in early spring before the new flower buds are differentiated: the period February to May is suitable in the temperate regions. But the most effective time of application does vary somewhat with the major elements, and it appears that nitrogen should be applied in late spring but not later than May. Potassium can also be applied in spring but phosphorus may best be applied in autumn. Finally, it appears certain that fertiliser application will in most cases have to be repeated if high seed production is to be sustained. Irrigation cannot be dealt with in detail here but it may have an important part to play in areas where water supply can become limiting to flowering and seed production.

Other Treatments of Trees and Ground

The stems of seed trees in older seed stands should be high-pruned to the base of the living crown to facilitate the use of the Swiss tree bicycle or special ladders for climbing to collect the seed, but no live branches should be removed from the base of the crown because this is the region of pollen production in many species. If truck-mounted ladders are used, any undergrowth and lop and top will either have to be removed from the site or stacked in heaps so that the truck can be maneuvered.

Strong weed growth often develops on the forest floor following the application of fertilisers and heavy thinnings; if allowed to grow unchecked, shrubs and weeds will be a serious barrier to the collection of the fruits and nuts of *Quercus, Fagus* and other species which are picked from the ground. Such weed growth can be suppressed by regular cutting and clearing or by the use of complete weedkillers such as "Simazine" or "Monuron" which have given promising results in apple orchards. Another possibility worth serious consideration is the use of leguminous ground crops; this has been the subject of much investigation for maintaining the health and yield of *Hevea* plantations in Malaya.

Protection against Fungi, Insects and Animals

The effects of these agencies in reducing flower, cone and fruit crops have already been stressed. The immediate application of creosote or other recommended chemical to the freshly cut stumps of felled trees in coniferous seed stands will reduce the spread of infection by *Fomes annosus*. Hardwood stumps which show signs of attack by *Armillaria mellea* should be grubbed up and burned where this is practicable.

When crown feeding insects assume epidemic proportions, as, for example, *Bupalus piniarius* on *Pinus sylvestris,* control should generally be achieved by applications of DDT, using ground fogging machines; aerial sprays are not normally economically justified. For tall trees in seed stands and smaller seed trees in seed orchards, systemic insecticides may eventually provide a suitable means of protection against the attacks of shoot borers and *Adelges* species, but this method is still at an early stage of development. Of the insects which attack flowers and seeds, *Megastigmus* species are among the most important at present.

The gray squirrel (*Sciurus carolinensis*) is a serious pest feeding on nuts, buds and bark. It can be controlled by poking out drays and shooting or trapping. Wide metal bands placed round the stems of widely spaced seed trees have been used in the United States of America to prevent red squirrels climbing.

Many species of both wild and game birds eat fallen seeds and some, for example the crossbill (*Loxia curvirostra*), extract seeds from cones. Control measures may be necessary and will depend on the wild life laws of the country concerned.

TREATMENT OF SEED ORCHARDS

Zobel *et al.* have defined a seed orchard as a plantation of genetically improved trees, isolated to reduce pollination from genetically inferior outside sources and intensively managed to produce frequent, abundant and easily harvested seed crops. It is established by setting out clones or seedling progeny of trees selected for the desired characters. Seed orchards are often established while genetic evaluation of the parent material is still under way. In such cases components shown by progeny testing to be genetically undesirable will subsequently be removed from the seed orchards.

Kinds of Seed Orchard

There are several kinds of seed orchard. Clonal seed orchards consist of grafts, rooted cuttings or layers; while seedling seed orchards are composed of selected progenies derived from open-pollination or controlled crosses. Clonal seed orchards may be subdivided into those in which the clones are from plus or untested parent trees; and those composed of clones from elite or tested parent trees. Concentrating first on clonal seed orchards based on plus trees, three kinds may be recognised on the basis of the origin of the clones:

1. Plus trees are selected from a specified geographic or climatic region; or from a group of stands; or, more rarely, from a single stand. All the plus trees are of the same species.
2. Plus trees are of the same species but originate from different geographic regions. These are called "provenance seed orchards" by Andersson who described one such seed orchard containing Swedish and Polish clones of Picea abies.
3. Plus trees are of different species and the object is to produce interspecific hybrids. Examples are : the seed orchards planted to produce seed of the F_1 hybrid *Larix decidua* × *L. leptolepis* = *Larix* × *eurolepis.* Andersson calls these "species crossing seed orchards."

Clonal seed orchards composed of elite or tested clones will form the next generation of seed orchards. Their formation is dependent on systematic test crosses made in the seed orchards composed of plus or untested clones.

Seedling seed orchards have been advocated by Wright who bases his argument on a theoretical comparison of the response to selection or genetic gain to be expected from several alternative selection procedures.

Choice of Site and Ground Treatments

Seed orchards are planted on sites well isolated from large areas of plantations of the same or related species, the distances recommended being up to 1,000 meters (1,100 yards), and favorable to the regular initiation of flower buds and the regular production of large crops of well-filled viable seed. The local climatic requirements are shelter from strong wind and adequate air drainage to reduce the damage done to flowers and fruits by frosts.

There are three possible ground treatments during the early growth of seed orchards: clean cultivation (this being the treatment favored in Sweden at the present time), cover crops and sward. Once the seed trees are well established, the sward appears to be a convenient treatment for many temperate tree species because of the relatively simple management and, by analogy with fruit trees, beneficial effects on the potassium and phosphorus status of the seed trees. On the three main types of sward used in fruit orchards, that is, the "tumble down," white clover sward and grass-white clover sward, the last appears to satisfy the three conditions of contributing soil humus, supplying some nitrogen (through the clover) and providing good ground cover. It should be noted, however, that drought effects can develop under *uncut* sward conditions so regular grass cutting or grazing is essential. Again, as with seed stands, grass swards are not suitable for all species and leguminous cover crops are worth consideration.

Manurial regimes for seed orchards are unlikely to differ much from those already suggested for seed stands, with the possible exception of methods of application. In top-fruit orchards (that is, apples, pears, etc.) certain fertilisers, in particular nitrogen and magnesium, are often more readily absorbed by the leaves of trees than through the roots. Olroyd made special reference to foliar applications of urea to top fruits.

Distance between Seed Trees

The most suitable distance between seed trees varies with species and methods of management, but initial planting distances should not be too wide because pollen production will be sparse and consequently the yield of viable seed will be rather low in the early years. Moreover, at a later stage, clones or seedlings shown by test crosses and progeny tests to be genetically inferior are also removed from the seed orchard, thus giving more growing space for the trees which remain.

Layouts designed to allow one or two "mechanical" thinnings are now general. The distances recommended range from initial spacings of 3 meters (10 feet) for *Pinus sylvestris* and *Fagus sylvatica* to final spacings of 6 to 9 meters (20 to 30 feet) for species of southern American pines. The arrangement of the clones of grafted plants or seedling progenies must meet the requirements of adequate cross-pollination and reduced self-pollination and so a randomised distribution of single seed trees is general. The value of clone or tree banks as a means of preliminary screening of clones for time and pattern of flowering, self compatibility and other related characters is worth stressing here.

Treatment of Seed Trees

In seed orchards the seedlings, grafts or cuttings should be encouraged by careful choice of site and favorable treatment at planting to grow rapidly from the beginning; this should lead to an early onset of flowering and fruiting. Thereafter, everything should be done to favor rapid growth and the best treatments appear to be regular crown release, the application of organic and inorganic fertilisers, control of the ground cover, pruning and geotropic treatments and, in some areas, irrigation. It may also be necessary to stimulate flowering by excessive applications of fertilisers, restricting water supply, stem girdling or root pruning, but in general these should be regarded as emergency or experimental, rather than regular, treatments. In the long term, choice of rootstock and treatment with plant growth substances will play their part.

Shoot Bending, Pruning and other Crown Treatments

The literature on these subjects has recently been reviewed by Matthews. The orientation of the shoot has a marked effect on flower initiation in apple and species of *Larix*. The geotropic or branch bending treatment developed for *Larix* by Longman and Wareing can be applied in larch seed orchards by tying branches of the seed trees into a downward pointing position during the dormant season. Goddard *et al.* report favorable flowering responses from bending over grafts of *Pinus elliottii* in Florida. Disbudding to induce or increase flowering has proved useful in *Pinus sylvestris*, and pinching off the shoot tips before meiosis takes place has been used in Japan in experiments on the artificial control of sex differentiation in *Pinus densiflora* and *P. thunbergii* . There is also scope for much more work on the effects of foliar application of growth regulators and nutrients on the initiation and differentiation of flower buds. Finally, it is clear that attention should in general be focused on the use of shoot bending, pruning, nutrients and growth substances to increase pollen production, which is often insufficient in young seed orchards of Pinus species.

Use of Selected Rootstocks in Grafting

It is now commonplace to find that grafted plants, formed by grafting scions taken from adult flowering trees on to seedling rootstocks in the juvenile nonflowering condition, soon begin to flower and bear seed. This is the basis of seed production in clonal seed orchards. In fruit orchards, selected clonal rootstocks are used to control the size of the grafted tree and the nature of the fruit bearing, and it would appear that the use of selected clonal rootstocks will also lead to similar benefits or uniformity and control of tree behavior in tree seed orchards. The idea is not new, Johnsson and others having grafted *Pinus sylvestris* scions on to seedling rootstocks of *Pinus mugo* and two origins of *Pinus sylvestris*. Analysis of the cone and seed harvest of 1959 and 1960 from these trees failed to reveal any influence of the rootstocks on the seed production of the scions, so the only evidence so far is negative. Nevertheless, an additional problem associated with the interaction of rootstock and scion is that of incompatibility of the graft union and this has been reported for *Pseudotsuga taxifolia* and *Fagus sylvatica* and may make necessary increased attention to selection of rootstocks.

Protection against Fungi, Insects and Animals

A factor which may have to be considered in the siting and subsequent treatment of *Pinus* seed orchards is the damage done by cone rusts. An example is provided by *Cronartium strobilinum* which damages the cones of *Pinus elliottii* and *P. palustris*. *Quercus virginiana* appears to be the most important alternate host, and Maloy and Matthews have recommended the siting of *Pinus elliottii* seed orchards away from *Quercus virginiana* and rust control measures in established seed orchards. In young seed orchards of *Larix*, *Pinus* and *Pseudotsuga taxifolia* the *Adelges* species which commonly appear on the grafted plants can be controlled by "Malathion" sprays. As in seed stands, systemic insecticides may also prove useful as a control measure in the future.

The Future for Seed Orchards

In conclusion, it may be said that the early development of seed orchards in Europe and America has been very promising, but the problem of mass-controlled pollination requires attention and more must be learned about the variation in self-incompatibility of the different clones and the role and nature of inbreeding. The early estimates of annual seed yields for *Pinus sylvestris*, which ranged from 11 to 16 kilograms of seed per hectare (10 to 15 pounds per acre), now appear, after 10 years' experience, to be sound, and the estimates for other species—for *Larix* 17 to 22 kilograms per hectare (16 to 20 pounds per acre) and for *Pinus taeda* 22 to 45 kilograms per hectare (20 to 41 pounds per acre)—also appear to be reasonable. There are now very considerable areas of seed orchard and, although most of them have been planted during the past 10 years, there is no doubt that they form a valuable and essential part of forest genetics and tree breeding work.

CERTIFICATION OF FOREST TREE SEED AND PLANTS

The object of the certification of tree seed and plants is to maintain and make available to the practicing forester sources of seeds, plants and other propagating materials of superior provenances and cultivars so grown and distributed as to insure the genetic identity and high quality of the seed and plants.

Superior provenances are identified by means of provenance tests. The duration of these tests depends on the rate of growth and the length of the economic rotation for each species but, in general, it must be accepted that provenance tests are essentially long-term experiments. The need to provide safe prescriptions for the movement of seed and plants so that expensive losses in wood production can be avoided has led foresters in countries with large variations of site and climate to devise zoning schemes defined in terms of latitude or distance and elevation. Thus in Austria, Czechoslovakia, Norway, Sweden and the Federal Republic of Germany to name a few such countries, seed is collected and the plants are utilised within the same zone. Countries with smaller ranges of site and climate tend, when importing seed, to look for seed sources growing in places with similar features of site and climate, although as the results of provenance tests accrue the specification can sometimes be made less strict and the search for superior provenances tends to range farther afield.

The state forest services of many countries maintain registers of seed sources which list classified seed sources for home use and export, and acceptable provenances that can be imported. General lists of acceptable provenances covering a wide range of species have been published in the United Kingdom, Denmark and Belgium and more restricted lists are available in other countries. Even if the best provenances are not yet known, foresters are often clear about what they should not use.

The rapid progress made with selection and breeding in the genus *Populus* has been largely due to the easy propagation of many species and hybrids by means of stem cuttings. There is a very wide range of distinct and stable cultivars in use, and control of genetic identity is exercised in most western European countries, partly by control of the sales of planting stock (as in Belgium, the Netherlands and Spain) and partly by the distribution of cuttings from registered stool beds.

Four broad categories of tree seed and plants are moving in national and international trade and these categories are distinguished by the precision with which origin is described; the degree of selection of the seed trees and their isolation from foreign pollen; whether the offspring of the seed trees have been tested or not; the amount of supervision and independent inspection of seed collection, processing and storage, and plant production; and the standards of record keeping and labeling. It is convenient to describe these four categories of seed and plants in ascending order of genetic value.

1. *Unclassified seed and plants.* The origin of this category is unknown or inadequately described, or the seed has been collected in stands not classed as seed sources by a recognised agency because of poor quality of the trees or lack of isolation from inferior trees of the same or closely related species. Unclassified seed is largely collected from felled or stunted trees and the seed, and plants are collected, raised and marketed with little or no supervision. The object of every country is to discard this category of seed and plants as rapidly as possible.
2. *Source-identified seed and plants.* This category is derived from good natural stands and plantations registered as seed sources by a competent service. The place or region where the seed is collected is clearly defined;

the seed is harvested, processed and stored and the plants are raised under supervision and independent inspection; the labeling is adequate and records are available for inspection.

3. *Selected seed and plants.* The seed is collected from carefully selected seed trees growing in natural stands and plantations classified as better than average and registered by a competent service. Trees that do not conform to stated standards of growth rate, stem and crown form, branching habit, health and other characters are removed as rapidly as possible from within the stand and from an isolation surround of prescribed width. These seed stands or seed production areas are also treated to increase seed production by means of fertiliser applications and protective measures against the animals, insects and fungi that damage seed. The statements made for source-identified seed about supervision, independent inspection, labeling and records also apply to selected seed.
4. *Certified seed and plants.* This category is derived mainly from clonal seed trees in seed orchards (and also from elite trees and stands) whose genetic superiority has been proved by progeny testing to standards defined by a competent service: Certified seed and plants are produced in a manner that insures genetic identity and proved interspecific hybrids may be included in this category. Clonal seed orchards contain highly selected clones arranged to favor cross-pollination. The site is well isolated from pollen of inferior trees of the same or closely related species.

Administrative Control of Seed and Plant Production

It is believed that there are 12 comprehensive national certification schemes and these can be grouped into 3 classes: voluntary schemes operated by independent seed associations; voluntary schemes operated by state forest services; and compulsory schemes operated by state forest services. The Czechoslovakian and Polish procedures are based on virtual state monopoly of tree seed and plant production, both for internal use and export. In the other countries with certification schemes, particularly the United States and central and northern Europe, there is a strong private seed and nursery trade. In the Federal Republic of Germany the Federal Seed Act of 1957 provides legal backing for a full official certification service. The scheme used in Georgia, U.S.A., is backed by two state laws but is voluntary; it is based on widely accepted procedures sponsored in the United States by the International Crop Improvement Association. In Finland and the Netherlands, there are private tree breeding associations that gain a good part of their revenue from the sale of certified seed and plants.

Those certifying agencies operated by independent associations are directed by a seed board or management committee elected by and representing the seed and nursery trade, forest owners, the state forest service and forest research institutes (including the seed testing agency). An important principle is that the purchasers of seed and plants should be represented. The management committee decides policy, sets up rules governing the certification program and has a professional forester as executive secretary or manager.

The voluntary certification agencies operated by state forest services usually have an officially-appointed supervisor who is responsible for the certification program under the direction of an appointed board that determines policy and sets up rules. Compulsory certification schemes are based on seed laws and are operated by the state agencies named under the seed laws.

Most certification schemes are nonprofit-making and largely financed by fees paid by the seed producers and nursery trade. These fees usually include a forest or nursery charge and an area inspection fee. Additional charges may be levied by means of official stamps or labels provided to the merchant by the certifying agency at a predetermined unit cost. The selling price of seed and plants of the superior categories is increased, often considerably, but the cost of seed is a very small fraction of total establishment costs and it can be shown that small increases in yield justify large increases in seed prices.

An Outline National Certification

The elements of a comprehensive national certification scheme for forest seed and plants are as follows:

1. *Inspection of the seed source by a qualified professional forester before pollination of the seed crop.* At this time (which is from 6 to 21 months before seed collection) the quality of the seed trees, incidence of inferior trees and effectiveness of the isolation can be checked. Many forest trees do not produce seed every year and, after the first inspection, subsequent re-inspections are made only in years when seed is to be collected (the longest interval being about 5 years).
2. *Assessment, if possible, of the cone or fruit crop by a qualified professional forester at a stated time (about 90 days) before collection begins.*
3. *Collection of the cones of fruits by a registered seed collector; extraction, cleaning and packaging of the seed at a registered seed extraction plant; and storage of the seed at a registered seed store.* The records of collection, processing and storage are made available for inspection and the labels must conform to minimum requirements.
4. *Testing, under the rules of the International Seed Testing Association, of an adequate sample of the seed at an official seed testing station.*
5. *Sowing of the seed in a registered nursery where labels and records satisfy minimum*

requirements. Inspection of the seedlings and transplants should be made by a qualified professional forester before they are lifted and dispatched.

Although growing-on tests are not yet widely used, the progress of research in several countries holds promise of their eventual use on a wide scale.

Minimum requirements for certification of seed and plants

The prescriptions that follow have been derived from the published accounts of national certification schemes in 12 countries. The prescriptions also incorporate relevant details from the scheme for the varietal certification of herbage seed sponsored by the Organisation for Economic Co-operation and Development.

Previous history of the seed source

A seed stand or seed production area may be formed in a natural stand (established by natural regeneration) or plantation (established by planting or sowing) of indigenous, introduced or unknown origin.

Isolation from foreign pollen

The requirements for isolation of seed production areas and seed orchards from large stands of inferior trees of the same or closely related species and for the width of isolation strips vary somewhat with species and country, but a generalised prescription for seed production areas and seed orchards is: 1,000 meters (1,100 yards) from a large stand of inferior trees and 100 to 150 meters (110 to 165 yards) from the nearest inferior tree—thus fixing the width of the isolation strip.

Field inspection of seed trees, seed crop and plants

Inspection of seed sources, seed production areas and seed orchards is most often made by the state forest service or by qualified professional foresters nominated by the certifying agency. The standards for seed production areas are based on the number of vigorous, well-formed and healthy seed trees per hectare and removal of inferior trees. Inferior trees are marked by the inspector and if possible felled while he is in the area, but the rate at which inferior trees are removed is often determined by local silvicultural requirements and re-inspection may be necessary when the fellings are completed. During seed collection and processing, the cones, fruit and seed must be handled so as to prevent mixture and maintain identity. Similarly, mixture must be prevented and identity maintained in the seed store and when raising plants in the nursery.

Standards for germination and purity

The methods of determining purity, germination and moisture content and the standards of health supervision with regard to seed-borne pests and diseases are generally based on those set by the Forest Tree Seeds Committee of the International Seed Testing Association (ISTA) Standards of germination and purity for purposes of sowing the seed and getting a good yield of plants are available for a very wide range of species.

Labels and certificates

It is generally agreed that the amount of information required on labels should be kept within reasonable bounds, and a number of distinctive yet simple labels are in use. Most of these bear the following items of information:

1. Species (Latin name)
2. Subspecies variety or cultivar name
3. Category (source-identified, selected, certified)
4. Provenance (reference number, place or region, elevation)
5. Supplier (name and address)
6. Seed (month and year of crop, quantity)
7. Plants (age or size, quantity).

In 1950, FAO and IUFRO jointly prepared an international certificate of origin and quality for tree seed and plants and this carries a large amount of information. This certificate has great value in exchanges of small quantities of seed for provenance research, but something simpler is needed for bulk trading in tree seed and plants. Such a certificate would bear the following items:

1. Certifying agency
2. Consignee (name and address)
3. Sender (name and address)
4. Species (latin name)
5. Subspecies variety or cultivar name
6. Category (source-identified, selected, certified)
7. Provenance (reference number, place or region, elevation)
8. Gross weight of packages
9. Contents (seed or plants, quantity)
10. Disinfection treatment (date, place, method).

Reference numbers

Comprehensive yet simple methods of identifying the origin of seed and plants by means of reference numbers or letters are of great assistance in administering a comprehensive certification scheme. FAO has adopted the universal decimal classification system by which each country is identified by a permanent number consisting of three digits.

Certification of forest tree seed and plants moving in international trade

Rohmeder points out that tree seed moves in national or international trade for three main reasons:

1. in large quantities for large-scale cultivation of a species, provenance or cultivar, as, for example, in *Pseudotsuga taxifolia, Pinus radiata, Tectona grandis* and *Eucalyptus;*
2. in small quantities for research purposes, such as species trials, provenance tests, progeny tests and tests of cultivars produced by selection and breeding;
3. as an item of food, such as chestnuts (*Castanea sativa*) walnuts (*Juglans regia*) or olives (*Olea europaea*).

The discussion that follows deals entirely with the first situation and is intended to refer to both seed and plants.

In Europe there has been regular and extensive international trade in seed and plants of *Pinus sylvestris, Picea babies, Larix decidua* and *Abies alba* for a very long time. Now there is extensive trading in tree seed between North America and Europe in the coniferous and broadleaved trees of the northern temperate zone, important examples being *Pinus contorta, Pinus nigra, Picea sitchensis* and *Pseudotsuga taxifolia.* Looking further afield, the *Eucalyptus* species and subtropical pines are playing a rapidly increasing part in afforestation in the Mediterranean region, central and southern Africa, South America, and the Asia and Pacific region. Finally, two tropical species deserve mention, namely the Honduras mahogany (*Swietenia macrophylla*) and teak (*Tectona grandis*), seed of which is moving extensively between tropical countries of Africa and Asia.

Six international bodies are active in matters affecting trading in forest tree seed and plants:

1. The Forest Seeds Committee of the International Seed Testing Association have developed rules for testing the quality of tree seed.
2. Section 22 of the International Union of Forest Research Organisations continues to sponsor international provenance research.
3. FAO has consistently sponsored improved records for use in the international exchange of tree seed, and organised the World Consultation on Forest Genetics and Tree Improvement to foster the use of genetically superior tree seed and plants.
4. The International Crop Improvement Association in the United States

has supported specific standards for certifying the origin of forest tree seed, drawn up by the Society of American Foresters.

5. Foresters of countries in the European Economic Community have agreed rules for trading in forest tree seed and plants.
6. The Organisation for Economic Co-operation and Development (OECD) has sponsored schemes for the certification of agricultural seeds.

The work of these international organisations suggests that the time is coming when internationally acceptable seed certification standards can be framed for forest tree seed and plants. The first steps have recently been taken by OECD which has provided funds to enable a working group to prepare the necessary draft rules and directives for consideration by all member countries.

REFERENCES

Allen, G. S. 2020. A method of distinguishing coastal from interior Douglas fir seed. *Brit. Columbia Lumberman,* 44 (8): 26-30.

Devitt, B. 2020. Fertilisation of two improved seed production areas. *For. Res. Rev. British Columbian For. Service,* p. 52-53.

Ehremberg, C., Gustafsson, A., Plym Forshell, C. P. & Simak, M. 2019. Seed quality and the principles of forest genetics. *Hereditas,* 41: 291-366.

Faulkner, R. 2022. Seed stands in Britain and their management. *Quart. J For.,* 56 (1): 1-15.

Georgia Crop Improvement Association, 2018. *Certification standards for forest tree seed (1961 crop, revised).* 5 p.

Hagner, S. & Bergman, F. 2021. On insamling av tallkott från stående träd. [Collection of pine cones from standing trees]. *Norrlands Skogsv. Förb. Tidskr.,* 1: 15-391.

International Crop Improvement Association. 2019. *Minimum seed certification standards.* Publ. No. 19. International Crop Improvement Association. 111 p.

Paged. 2022. *Forest tree seeds and trees.* Warsaw, Paged Foreign Trade Enterprise Export and Import.

10

Seed Processing

The objective of fruit or seed processing is to achieve clean, pure seeds of high physiological quality (germinability) which can be stored and easily handled during succeeding processes, such as pretreatment, transport and sowing. Processing includes a number of handling procedures, where applicability differs e.g. according to fruit and seed type, condition of the fruits or seeds at collection and potential storage period. Processing can be grouped into the following 7 procedures:

1. Pre-cleaning, for fruit or seed lots containing larger debris, leaves, twigs, empty fruit parts etc.
2. Pre-curing, for fruits that must be after-ripened, or where rapid desiccation hampers extraction.
3. Extraction, for species where the fruits are collected but only the seeds (and occasionally part of the fruit) are stored and sown.
4. Dewinging, for fruits and seeds with wings. Also including removal of dry appendices like spines, arils and hairs.
5. Cleaning, for fruits or seeds with impurities like fruit parts, leaves, twigs, empty seeds, foreign seeds and chaff.
6. Grading, for seed lots with large variation in seed size or weight.
7. Adjustment of moisture content, for seeds which, after the other procedures, have a higher or lower moisture content than considered optimal for storage of the particular species for the expected storage period.

Seed processing normally follows this order, but certain steps may be irrelevant and hence omitted for particular species or seed lots.

Any step in the processing procedure must be carefully adjusted to the particular fruit or seed type. Processing implies a risk of losing seeds both by under and over-treatment. Under-treatment may fail to achieve the desired result (e.g. insufficient extraction of seeds); over-treatment may damage the seeds with consequent loss of viability or reduced storability. Both under and over-treatment may influence the genetic composition of the seed lot.

Processing should, as far as possible, take place immediately after the fruits or seeds have been brought to the processing depot. Yet, temporary storage can rarely be avoided since the processing facilities are often busy during peak collection season. The timely order of processing should give high priority to species and seed lots with easily deteriorating seeds, e.g. fleshy fruits and recalcitrant seeds, or seeds collected with a high moisture content. Processing is not necessarily one continuous process. Partial or intermediate processing, like pre-cleaning, drying or removal of fleshy fruit parts, may typically be undertaken upon arrival to the processing depot in order to impede deterioration during transit. Final processing may then be postponed until later.

Several types of mechanical equipment are available for seed processing. Typically, forest seed processing equipment (as different from agricultural equipment) must be able to adjust to many different seed types. Several types of mainly dry fruits develop large amounts of dust when processed mechanically. In addition to being highly· unpleasant dust can cause irritation of skin and respiratory system. Safety precautions and protection of labourers are essential during processing of these species.

PRE-CLEANING

As during field operations it is essential that equipment and containers are clean, i.e. free from pathogens and seeds of other species, in order to avoid any contamination and possible infection, and that seed lot identity is carefully maintained throughout the operation.

Pre-cleaning is the removal of larger matter such as leaves, twigs and empty fruits. In the field pre-cleaning is undertaken primarily to reduce bulk during transport and storage. If bulk reduction has not been undertaken in the field, it may be relevant at the processing depot. Redundant bulk inevitably hampers the efficient use of processing equipment, simply because it adds to the volume to be processed.

In some cases pre-cleaning may be necessary to remove material that may impede efficient extraction and subsequent cleaning. For ex-ample, twigs left with acacia pods to be threshed for seed extraction break down to size and weight similar to the seed during threshing. The twig fragments are very difficult to separate from the seed during the subsequent cleaning operation. In casuarinas the cladodes (scaly leaves) readily fall off the branches when dry and since they are about the same size as the seeds, they make seed cleaning difficult, time consuming and hence expensive. The fruits must therefore be stripped off the branches before extraction.

Finally, leaves, twigs or soil particles may carry fungal spores or other pathogens, which may infect the extracted seeds or germinating seeds and seedlings. An example of the latter is the needle cast disease that affects certain pines. Fungal spores of this pathogen are carried on the dry needles and may from there contaminate the seed lot. The sooner the needles are removed from the seed, the easier it is to control the disease.

Pre-cleaning is usually done manually after arrival at the seed-processing depot. Large quantities of fruits are pre-cleaned mechanically e.g. on vibrating or oscillating screens or in tumblers. The design of pre-cleaning equipment depends on fruit and seed type and the character of debris (twigs, leaves, stones, soil etc.).

PRE-CURING

Pre-curing denotes the procedure during which fruits are kept moist for a prolonged period before extraction. Pre-curing has two ration-ales: 1) to promote after-ripening of immature fruits, and 2) to ease extraction of seed where rapid desiccation may cause extraction problems, in extreme cases case-hardening.

Seeds, which deliberately or incidentally were collected immature, must be after-ripened to become physiologically germinable. The critical stage after which after-ripening is possible varies between species and must be based on experience. A few species like Ginkgo biloba and temperate Fraxinus species always require after-ripening since the embryo of their seeds is underdeveloped at the stage of dispersal.

During the natural maturation process, the water supply to the maturing fruit is regulated through the pedicel and to the seed through the funicle. Continuous evaporation prevents overheating of the embryo while it still has a high moisture content. Maintenance of high moisture level and avoidance of drastic increase in temperature should simulate these conditions during after-ripening.

Only fruits that are not mature should be after-ripened. If the seed lot varies in maturity, as can usually be judged from the appearance of the fruits, the fruits should be sorted prior to processing: small under-developed fruits are always discarded as they have not attained the capacity to after-ripen. Fully mature fruits go directly to the next step in the processing chain, e.g. extraction; an intermediate portion may consist of mature size fruits capable of after-ripening during a pre-curing period. For example, in Malaysia pods of Acacia mangium are separated into three classes according to colour viz. greenish-brown, brown and black. Greenish-brown pods are after-ripened for 120 hours, brown pods for 72 hours, and black pods go directly to extraction by kiln drying. In Thailand pre-curing of Pinus merkusii and P. kesiya is routinely done by storing freshly collected cones in loosely tied gunny-bags or bamboo baskets for 7-14 days (depending on maturity) in a ventilated room or shed before extraction . In Kenya seeds of Azadirachta indica, Thevetia peruviana and Ximenia americana are after-ripened for 2-3 days after collection.

After-ripening typically takes from a few days to a few weeks. The environment during this period is important in order to control the physiological processes. The fruits are kept at normal air temperature, in the tropics between 20 and 30°C. The moisture level is initially high but gradually reduced during the process. The high moisture level during pre-curing makes the seed susceptible to fungal attack. Proper ventilation limits that risk, but carries a risk of desiccation.

After-ripening is concluded when the fruits have reached full maturity. In most dry fruit types the maturity indices of naturally matured fruits will hold for after-ripened fruits as well. However, since after-ripened fruits are not normally exposed to sunlight, maturity colours of fleshy fruits may be slightly different from naturally ripe ones, and the fruit flesh may remain relatively firm. To judge when maturity has been reached, examination of seed development and the embryo may be carried out by a cutting test.

Pre-curing of cones primarily serves the purpose of easing extraction. In several conifers rapid drying causes insufficient expansion of the cone scales with consequent difficulties in extracting the seeds. 'Case-hardening' is the situation where moisture is trapped inside the cone because the outer layer of the cone dries so quickly that the cells 'collapse' and prevent moisture from the interior migrating to the surface. This may also occur if the cones are not given enough space to expand during drying, for instance when they are tightly packed in bags or containers. Re-moistening the fruits, and then exposing them to a second slower drying process normally overcomes the problem.

Seed extraction of several tropical pines can be greatly facilitated by pre-curing, especially for early collected cones. In an experiment of early-collected cones of Pinus elliottii the number of seeds extracted per cone increased from 0 to 60 for cones pre-cured for 1 and 5 weeks respectively. When cones were collected one month later, the number of seeds extracted increased from an average of 27 after 1 week's pre-curing to 82 after 5 weeks' pre-curing.

Pre-curing of relatively dry fruits may be conducted while the fruits are stored in containers or bags. Alternatively, the fruits may be spread in a thin layer (one fruit thick of large fruits) on concrete floors or in trays. In the latter case it is easier to control the environment during the pre-curing period. Desiccation may be controlled by regularly spraying with water. Spraying is gradually reduced as the fruits reach full maturity.

Pre-curing of fleshy fruits should always be carried out while the fruits are spread out. To allow moisture from softening fruits to drain off, the fruits should preferably be pre-cured in open trays. If the fruits attain maturity at different rates, it may be necessary to manually remove mature fruits in order to avoid decomposition or fermentation of the pulp while pre-curing is still going on.

The process of pre-curing is basically the same whether used for after-ripening or to ease extraction. However, since the latter is a mainly physical process, it is less dependent on a physiologically optimal temperature. Consequently the temperature regime is less critical. For example, for cones it may be advantageous to gradually raise the temperature during the later stages of the process.

SEED EXTRACTION

Extraction denotes the procedure of physically releasing and separating the seeds from their enclosing fruit structure.

The main rationale of extraction is to:

1. *Reduce bulk:* The seeds typically make up 1-5% of the total fruit volume. Bulk reduction helps to reduce cost of storing and shipment.
2. *Ease handling:* Seeds are normally tested, pretreated and sown individually, which makes their separation from the fruit necessary.
3. *Improve storability:* Easily decomposable fruit parts such as the pulp of fleshy fruits or arils must be removed to avoid their decomposition during storage. Moisture contained in dry fruit types and cones may attract fungi and insects, especially if stored under ambient temperature. In addition, drying of seeds to safe moisture content becomes difficult if they are not extracted.

Seeds are extracted from most multiple-seeded dry fruits such as capsules, pods and follicles and from most fleshy fruit types. In single or few-seeded dry fruits like samaras and nuts, the whole fruit is normally stored and sown, albeit sometimes after removal of e.g. wings or other external structures. In a few fleshy fruits like Vitex parviflora, and arillate seeds of some Podocarpus spp. the whole fleshy unit may be dried and stored. In drupes extraction is only partial as only the fleshy pulp is normally removed, the seed remaining within the dry endocarp. In multiple or compound fruits like Casuarina, Morus and Banksia, extraction is also only partial in the sense that the single-seeded fruits (samaras, achenes or nuts) are extracted from the compound bulky fruit structure.

Extraction is usually undertaken prior to storage, but in some species it may be delayed until just before sowing or omitted altogether, e.g.:

1. Where storability of non-extracted seeds is considerably better than that of extracted seeds.
2. Where labour requirement for extraction is so high that it outweighs possible gain from extraction, e.g. ease of handling and germinability.

3. Where storage facilities, shipment, transport or other handling procedures do not make extraction mandatory.

Reduced storability of extracted seeds may be due to the absence of the protecting structure of the fruit. Further, extraction of seeds of some species is so difficult that it almost inevitably causes physical damage to the seed. Such damage is likely to affect storability. For example, Cedrus seeds do not store well when extracted and the seeds are therefore normally stored within their fruits, and extracted just before sowing. In Melia volkensii the normal storage unit is the depulped drupe (pyrene). The pyrene is very hard and contains 3-4 seeds. In order to extract the seed, the endocarp must be broken, a very arduous manual procedure that inevitably damages some seed. If the damage does not affect delicate parts of the embryo (e.g. the radicle), the seeds may germinate normally if sown immediately. However, even lesser damage is likely to reduce their storability. In both examples the desired bulk reduction by extraction should be balanced against a likely reduced storability.

Table 1. Examples of different storage units in different species

Seeds normally extracted before storage	Most species, e.g. most conifers, casua-rinas, eucalypts, and species of the families Leguminoseae, Meliaceae, Bignoniaceae.
Seeds normally stored and sown together with the entire fruit	African *Terminalia* spp., *Quercus* spp. and dipterocarps
Seeds often or occasionally stored within fruits but extracted before sowing	*Vitex parviflora, Maesopsis eminiii, Grewia* spp., *Pterocarpus* spp

In addition to the possible physical damage to the seed, the process of extraction is often very labour intensive. For samaras like Terminalia brownii and several Pterocarpus species, and for stones of e.g. Melia azedarach no effective extraction methods exist. Despite their bulkiness and possible delayed germination, seeds of many hard fruited species are not fully extracted but stored and sown together with the enclosing fruit.

Table 2. Summary of extraction meth-ods for various fruit types

Fruit type	Extraction procedure
Dry dehiscent fruits, e.g. dehiscent pods, follicles and capsules and cones, e.g. pines, eucalypts and most Leguminosae	Drying → shaking / tumbling
Dry indehiscent fruits, e.g. indehiscent pods of *Acacia nilotica* and *A. siberiana*	Drying → threshing
Serotinous fruits, e.g. cones, capsules plus some dry compound fruits	Kiln heating → tumbling Scorching → tumbling
Fleshy fruits w/ very thin pulp, e.g. *Vitex* spp. and *Ziziphus* spp.	Drying Soaking → maceration → washing
Fleshy fruits w/ soft pulp, e.g. *Prunus, Olea, Ficus*	Soaking → fermentation → washing Soaking → maceration → washing
Fleshy fruits w/ soft, fibrous pulp, e.g. *Gmelina*	Soaking → maceration → washing → abrasion / polishing
Fleshy fruits w/ felty pulp, e.g. *Tectona grandis, Sclerocarya* spp. and *Vitex* spp.	Soaking → abrasion / polishing

Extraction may, as in the case of Vitex parviflora, Melia volkensii and Pterocarpus spp., be carried out just prior to sowing. Since the purpose of such post-storage extraction is to promote germination, it is in seed handling context equivalent to pretreatment.

Most fruit types can be classified as dry or fleshy. During extraction the former is dried to low moisture content, the latter extracted moist, often after initial soaking in water. A summary of extraction procedures appears from table 2. Some intermediate types may be extracted as either dry or wet or by two succeeding procedures. For example, some Prosopis spp. are first extracted from the pod by dry extraction i.e. drying and threshing; the enclosing pulp is then removed by washing.

In Afzelia africana the enclosing aril may be removed by washing after dry extraction of the seeds from the pod. In many-seeded drupes like Melia azedarach depulping is carried out according to the procedure for fleshy fruits, while an extraction procedure for dry fruit must be used if the seeds subsequently are extracted from the stone.

Dry Fruits

Morphologically, dry fruits are classified as dehiscent and indehiscent fruits. Dehiscent fruits open at maturity to release their seeds; indehiscent remain closed at maturity. This classification is, however, more complex than the definition suggests since there is a gradual transition from indehiscence to dehiscence, which has practical implications for seed extraction. Extraction from dehiscent fruits is influenced by:

1. *Fruit type:* Dehiscent fruit types are follicles, capsules, cones, dehiscent pods and some compound fruits. Dehiscence is generally related to species (e.g. dehiscent pods of Acacia mellifera, indehiscent pods of A. nilotica), but ease of extraction sometimes varies with subspecies, variety or provenance.
2. *Stage of maturity at collection:* Fruits picked fully mature are often more likely to dehisce than early collected ones. The difference may be fully or partly overcome by pre-curing.
3. *External environment:* Primarily the moisture content in the fruit determines dehiscence; when the fruit dries out, it tends to split open. In some fruit types, primarily cones, the moisture content is in equilibrium with air humidity such that when the air is dry, the fruit loses moisture and splits open; when the air is humid the fruit may re-gain moisture and close again. The process is hence reversible.

Dry fruits are normally handled according to degree of dehiscence. Four groups are recognised.

(a) Dehiscent fruits that open upon drying

(b) Serotinous cones and fruits

(c) Indehiscent fruits where seeds are extracted

(d) Indehiscent fruits where seeds are not extracted

A. Dehiscent fruits that open upon drying

Many dehiscent fruits open readily when dry, and extraction of their seeds rarely imposes any problems under dry conditions. Examples of some easily extractable seeds are species of Eucalyptus, Callitris, Casuarina, and many conifers. Often seed extraction of these species can be done easily in the field. All that is needed is space to spread out the fruits in a thin layer and sufficient air circulation. Techniques recommended for drying in the field hold for the processing depot as well. Dry processing at the processing depot can, however, be provided with more permanent drying facilities, e.g. concrete platforms, shelters, drying trays and possibly artificial air circulation..

Most dehiscent cones open readily when their moisture content (m.c.) falls to 20-25% and are fully opened at a moisture content of 10-15%. For most species, 15% m.c. is equivalent to a relative air humidity (RH) of around 70%. Under humid conditions as in the ever-humid tropics or during the rainy season, RH rarely comes below 70% in the shade.

Under these conditions it can be a problem to bring down the m.c. of the fruits with concurrent dehiscence. Direct sun drying is an option that should be handled with care since temperatures in the sun may easily reach 45-60°C, a critically high level for seed with a high m.c.

Temperature can be raised with concurrent improvement in drying effect by covering e.g. extraction trays with transparent polythene sheets. The risk of detrimental overheating of moist seed is the same as for direct sun drying, and high temperatures should be applied only when the fruits have lost most of their moisture.

To avoid overheating, desiccation should be gradual. The fruits may be exposed to a maximum temperature of 35-40°C until the moisture content of the seeds has lowered to 15-20%. After that the temperature may be raised e.g. by sun or kiln drying. For immature and moist material of most Australian species, ATSC recommends an initial drying for two days at 25°C, then increasing to 35°C. The low initial temperature helps minimise the risk of fungal activity destroying the seeds.

Outdoor drying is feasible for many dry fruits during the dry season. The fruits may be dried on concrete platforms, tarpaulins or permanent drying shelters. Because of the reversible character of dehiscence, it is sometimes necessary to prevent the fruits or cones from re-absorbing moisture when the air humidity rises, e.g. at night. The fruits may be covered or moved indoor, whenever there is a risk of re-gaining moisture.

In cases where air humidity is too high or where frequent showers make sun drying impossible or impractical, the fruits must be kiln dried. Kiln drying is used for most temperate cones. In kilns the temperature is raised above ambient air temperature by solar or artificial heating. To assure uniform drying the fruits are regularly stirred or moved during kiln drying. Kiln drying may be combined with tumbling by providing the tumblers with a heated air stream.

The physical release of seeds from dehiscent fruits varies with species and condition of the fruit. Funicle attachment or physical trapping within the fruit may impede the complete detachment from the fruit. Capsules of *Swietenia* and *Entandophragma,* and cones of *Abies, Araucaria* and *Agathis* normally disintegrate completely upon drying. In some species a minor mechanical impact such as raking, shaking or tumbling is sufficient to complete extraction. That goes for e.g. *Markhamia platycalyx, Spathodea nilotica* and other *Bignoniaceae,* and many conifers. In eucalypts the release of seeds depends upon the position of the ovaries. Species with half-superior ovaries like E. ca-maldulensis release their seeds more easily than those with inferior ovaries, e.g. *E. delegatensis*. Chaff in the capsules can be a physical constraint for the release of seeds in eucalypts. Seeds of some species maintain a strong attachment to the fruits via the funicle after dehiscence. That goes especially for dehiscent legumes like *Acacia, Albizia, Acrocarpus* and *Paraserianthes,* which are naturally dispersed attached to half of the dehiscent pod. The mechanical release may be accomplished by tumbling or shaking, e.g. together with pre-cleaning of the seed. If the attachment is very strong, the seeds may need to be extracted by the methods of indehiscent fruits, e.g. by threshing or flailing.

Fruits of normally dehiscent species may fail to open if collected immature. For example, greenish and light brown pods of *Acacia mangium* often retain a significant number of seeds after pre-curing and threshing. Kiln drying may be able to open the fruits.

Capsules and cones attacked by insects or fungi may occasionally fail to open appropriately, either because web or fungal hyphae physically obstruct seed release, or because the infection obstructs the normal opening mechanism. Seeds from such fruits can only be extracted by threshing.

B. Serotinous cones and fruits

Serotinous cones and fruits are morphologically dehiscent, but their dehiscence requires an exceptionally high temperature, in nature encountered during bush fires. Serotinous cones occur in pine species like *Pinus taeda, P. brutia, P. halepensis* and *P. contorta*. In these species the cone scales remain closed due to sealing with resin. During exposure to high temperature the resin melts and the cones open by the same hygroscopic mechanism as other cones. Serotinous fruits also occur in some angiosperms from fire prone areas, e.g. some *Eucalyptus, Casuarina, Banksia* and *Hakea* species. Many serotinous cones and fruits only open upon exposure to temperatures of at least 70-80°C for several hours. Therefore seed extraction normally takes place in artificially heated kilns. Kiln drying at 70°C for 18 hours was found suitable and did not hamper viability for pre-cured Acacia mangium seeds.

A brief exposure of max 1.5 min to super-high temperature of 220°C provided by a gas flame was found a suitable alternative to prolonged kiln drying for extraction of seeds of *Pinus contorta* var. *latifolia*. Longer exposure seriously hampered viability.

In Australia fruits of *Banksia* and other extremely hard fruits are opened by placing them on a wire mess over hot coal until they split open. When the fruits have opened, they are immediately immersed into water and then sun dried. Short dipping in boiling water for 5 sec. to 10 min has been reported to ease opening during subsequent kiln drying for some species. Alternatively, where kiln facilities are not available or for less serotinous fruits, several cycles of drying and wetting at normal temperature usually help to open difficult dehiscent fruits.

Although species with serotinous fruits are adapted to high temperature, too high temperatures or prolonged exposure may be detrimental. It should be reiterated that high moisture content in the seeds makes them more sensitive to heat. Therefore, pre-drying of moist serotinous fruits should be undertaken before exposure to high temperature. This will also prevent the above-mentioned case-hardening.

C. Indehiscent fruits where seeds are extracted

Indehiscent pods of e.g. *Acacia* and *Prosopis spp.* will not open by drying alone, but must be broken mechanically in order to extract the seeds. Initial drying or heat treatment promotes brittleness of the pods, which will facilitate subsequent extraction. For Australian species drying at 40-45°C for 24 hours has been recommended. Seeds from large fruits like Parkia biglobosa may be extracted by individually splitting each fruit by hand. Most smaller indehiscent pods may be broken-up by beating or flailing, crushing them in a mortar, rolling over them with a drum, threshing or tumbling in a cement mixer with blocks of heavy wood. Any mechanical treatment that will split the fruits without damaging the seeds is suitable.

Most mechanical threshers work on the same principle. The thresher consists of a revolving beater mounted within a horizontal cylinder. Fruits are fed from one side and torn apart by the beater. Seed and small fruit parts may pass perforations in the bridge (the concave plate beneath the beater) or the sieve behind, while larger material is removed. The seeds are thus pre-cleaned to-gether with the mechanical extraction.

The most commonly used thresher type for forest seed is the flailing thresher, which is used e.g. for several Australian *Acacia* species. Mechanical threshers designed for agricultural crops are available in most parts of the world. They are normally designed for grass crops like rice or grain and usually need some modification to be used efficiently for woody material.

For species with very hard pods like *Acacia nilotica* and some *Prosopis* species, threshing is insufficient to break the pods. Grinding mills have been used successfully for some of these species. The grinding stones or steel are adjusted to a distance slightly larger than the seed. Hammer mills work in principle like a flailing thresher, where the beater is replaced by a revolving steel cross pulverising the material against the drum. It can be used for threshing hard fruits if the revolutions of the hammer are reduced to 250-800 per minute and the outlet screen is replaced by holes that will let the seeds out.

In species like several *Prosopis spp.* the seeds are embedded in a pulpy material, or the pod contains gummy material which is difficult to break up mechanically. These species must be treated in a series of steps:

(i) Breaking the pods

(ii) Soaking in 0.1N hypochloric acid for 24 hours

(iii) Washing and drying iv. Pounding the dry material with a hammer or in a mortar.

Pods with gummy material (e.g. Prosopis cineraria) may need several rounds of threshing with intermediate drying, or the pods may be run through a coarse meat grinder to extract the seed.

Seeds of indehiscent pods of acacias are adapted to being ingested by animals; their seed-coats are very hard and are rarely damaged by mechanical extraction. Other seeds are more delicate and me-chanical treatment should be limited as much as possible. Seeds that have been extracted are more susceptible to physical damage than those still retained within and protected by the fruit. It is important, therefore, that seeds are removed from the fruits soon after their release. In some instances 2-3 rounds of threshing with intermediate removal of extracted seeds are better than one complete threshing at high speed.

D. Indehiscent fruits where seeds are not extracted

Seeds in samaras of e.g. *Pterocarpus, Combretum* and *Terminalia* are not normally extracted but remain within the fruit at least until sowing. Nuts of *Quercus, Fagus* and *Castanea* are usually extracted from their involucre or cupula by drying or tumbling. Dipterocarps are not extracted.

Fully mature fruit pulp of several drupes, berries and other fleshy fruits often separates readily from the seeds. The pulp may be detached manually or by washing. Other species like *Azadirachta indica, Aleurites spp.*, and *Santalum spp.* have relatively firm pulp that need softening to ease seed extraction.

Softening occurs during decomposition of the pulp, normally as a result of fermentation. Controlled fermentation can be used effectively as a pre-extraction procedure for several species e.g. *Dovyalis*. However, heat and possibly alcohol produced by the process sometimes have a negative effect on seed quality. Seeds extracted from fermented *Gmelina arborea* drupes have shown a markedly reduced germinability as compared to seeds extracted mechanically from fresh fruits.

Soaking in water for one to several days often accelerates decomposition of the pulp, especially for fruit types with relatively dry pulp, e.g. *Zizi-phus mucronata* and some *Diospyros* species. Soaking for several days has been recommended for e.g. *Santalum spp.* the pulp of which is generally difficult to remove. Mechanical rupture of fruit skin prior to soaking often speeds up softening and decomposition. If the fruit pulp has only been partly removed during field handling, the remainder having been dried, the seeds normally need to be soaked or wetted again to remove the remaining pulp at the processing depot. Dur-ing prolonged soaking of several days, water must be changed regularly e.g. once every 12 hours, or the container with fruit must be provided with a continuous water flow to prevent fermentation damage.

After soaking or decomposition the pulp and fruit skin are separated from the seed. Several methods, often used in combination, are applicable. Selection of the most appropriate method depends on fruit type, quantity of fruits to be processed and equipment available:

1. *Individual manual extraction:* Entirely manual extraction may be applicable to very small seed lots, seeds with extremely fragile seed-coats (e.g. *Syzygium cumini*), or large fruits where mechanical equipment cannot be efficiently used. The latter category encompasses many humid tropical species like *Artocarpus and Durio* spp. Individually extracted seeds may be washed and cleaned.

2. *Washing in deep bowls or drums:* This may be a simple continuation of soaking and applicable if the pulp separates easily from the seeds or stones, *Melia volkensii, Dipteryx panamensis, Prunus africana* and others. Mechanical stirring or using a strong water stream while the fruits are submerged in water help to loosen the pulp. If pulp sticks to the seeds or stones, vigorous stirring or increased water pressure is necessary. Small seed lots and small seeds may be extracted effectively by the aid of a blender used at low speed or an electric mixer. Separation of pulp + skin and seeds is done by flotation in excess water: seeds or stones remain at the bottom while pulp and skin tend to ascend to the surface where they may be skimmed off. In some species very clean seeds can be achieved by washing, in others washing may be used as a pre-extraction procedure.

3. *Washing on wire mesh screens:* Screens with mesh size that will retain the seeds while the pulp passes through are used. The pulp is released by manually rubbing the fruit against the grid and washing. High water pressure should be avoided because of the risk of washing away seeds. Fruit skin and firm fibres of the fruit flesh are normally retained with the seeds or stones after rubbing and washing. They may be separated by flotation in excess water as described above. The method is widely applicable to most fleshy fruit species, and the safest and most efficient procedure for extraction of very small seeds or seeds with fragile seed-coats. Since the fruits are rubbed manually on the screen, the mechanical impact can easily be adjusted when extracting fragile seeds. To separate pulp from tiny seeds, fine mesh screens must be used. In the Philippines a 1/16 inch mesh was used for seed extraction of Anthocephalus cadamba. If high-pressure water is available, a modification of this method is to place the fruits in wire mesh bags with mesh size just below the seed or stone size. The water pressure is then directed directly against the fruits, which may be cleaned completely by the water pressure. This method has been used to clean seeds of e.g. Prunus and Vitis spp..
4. *Cement mixer with abrading material:* The fruits are mixed with an abrading material like gravel plus excess water and rotated in the drum for various lengths of time, typically from 5 to 20 min. While the abrading material rubs against the fruits, the pulp is gradually torn off. The fruits are checked regularly in order to avoid unnecessary overtreatment with possible damage to the seed-coat. Cleaning becomes more difficult with this method since the mixture must be separated into three fractions after extraction, viz. seed, abrading material and fruit pulp. The method may be used as the only depulping method; more often, however, it is used as a final cleaning after the major part of the pulp has been removed e.g. by washing. The risk of mechanical damage is relatively high and the method hence less applicable to seeds with fragile seed-coats. Tumbling in a cement mixer with blocks of wood has been used in Kenya for species with sticky pulp e.g. Vitex keniensis, Maesopsis eminii and Cordia spp..
5. Mechanical depulping. For large quantities of fruits or where the flesh tends to remain firm, mechanical depulping is applicable. In the tropics the most widely used mechanical equipment is the coffee depulper, or a modified version thereof. The depulper mechanically abrades the fruit pulp by rupturing and squeezing against or between its mechanical parts. Because of the risk of mechanical damage, the depulper is particularly applicable to fruits with relatively hard seed-coats or stones, e.g. most drupes. A slightly modified version of the ordinary coffee depulper has been used effectively for Gmelina arborea. Another

widely used mechanical depulper is the 'Dybvig' macerator. In this depulper the fruit pulp is abraded on a flat spinning plate provided with four bars arranged in a 90 degree cross at the bottom of a cylinder. A continuous water stream helps to wash away the separated fruit pulp. The standard model of the 'Dybvig' macerator has been improved by lining the inside of the cylinder with a wire net and bolting a can, also lined with wire net, to the spinning plate.

In some species, fibrous fruit flesh tends to remain attached to the endocarp or seed-coat after washing. These remnants may be removed by abrasion, polishing or brushing. Dry or moist tumbling with sand or other abrading material in a cement mixer is effective, but may damage sensitive seeds. Seeds of Gmelina arborea have been success-fully cleaned for residual pulp by polishing them in a coffee dehusker.

In teak (*Tectona grandis)* the mesocarp is felty rather than fleshy, and its removal requires a relatively vigorous treatment. In India a depulping machine with a drum tightly wrapped with barbed wire was found suitable.

Depulping of species with extremely sticky fruit flesh, e.g. some Euphorbiaceae like *Antiaris toxicaria* and *Bishofia javanica* may be facilitated by adding some detergent or alkaline to the water during tumbling, e.g. 1 N hypochlorite. Cleaning of equipment after extraction should be done with a similar detergent, but the concentration may be increased.

Germination of seeds in fleshy fruits may to some extent be prevented by inhibitory compounds in the fruit flesh. Once the fruit pulp has been removed, the seeds may be able to germinate if the moisture content is high. Therefore, orthodox seeds should be dried quickly. Recalcitrant seed should be dried to the lowest safe moisture level and stored at the lowest safe temperature to prevent germination.

Biological Extraction

Fleshy fruits and several indehiscent dry fruits are adapted to being ingested by animals. Seeds and stones are often left cleaned and intact after ingestion although in some species a relatively large amount of seeds may be digested.

Feeding legume pods to domestic goats may serve the purpose of extracting seeds from the fruits, but implies a subsequent workload in collecting the droppings and then extracting the seeds. The procedure can be rather laborious, but ingestion usually scarifies the seeds so that labour may be saved on pretreatment. Accumulated manure in goat enclosures in areas with heavily fruiting *Acacia nilotica* and *A. tortilis* often contains large amounts of seeds from these species.

Seeds can be extracted from manure by wet or dry extraction. During dry extraction the manure is initially dried and fractioned e.g. by gentle pounding in a mortar or the like, then cleaned by tumbling and sifting. During wet extraction the manure is soaked and washed in water. The seeds that gather at the bottom of the container are then separated by sifting under running water. Wet extraction gives the cleanest seed, but if scarified by the ingestion, the seeds may readily imbibe, which may make them sensitive to further treatment. A problem of collecting the seeds from manure is that the manure often contains a mixture of seeds from different species which can be difficult to separate.

Both ants and termites may readily and efficiently clean fruits and seeds of sweet material, both moist and dry. In the Philippines, termites will readily attack pods of *Samanea saman* piled up in a dark place. The termites consume the fruit parts only, and after a short time the seeds will be left clean. Termites also attack fruits of Kigelia (sausage tree), the seeds of which are very difficult to extract except manually. However, although it is primarily the soft fruit that is consumed, many seeds are also lost. In Brazil, ants have been reported to efficiently remove pulp from pods of *Hymenaea courbaril* (Caesalpinaceae) which also promoted germination of the seed. Termite extraction is generally not a good idea close to buildings and inside nursery premises. The indirect feeding may both attract and assist them to multiply, which may give problems later. Ants are less problematic and may be used for final cleaning of residual adherent pulp.

DEWINGING

De-winging, in a broad sense, is removal of any dry seed appendage, including wings, spines, hairs, and some aril types. Seed (or fruit) wings do not obstruct germination, but may be inconvenient in handling. Accordingly, the main purpose of dewinging is to reduce bulk and ease handling during storage, pretreatment and sowing. In some instances wings, hairs or other appendices, which increase the surface area of the seed, tend to collect moisture and promote fungal attack. Dewinging is often done as a routine before storage or shipment, while seeds sown immediately after collection, e.g. dipterocarps and other recalcitrant species, rarely have their wings removed.

Wings and hairs occur in wind dispersed species. Wings are typical of indehiscent dry fruits like samaras of *African Terminalia spp., Pterocarpus, Triplochiton, Heretiera, Kokoona, Casuarina* and many others. Winged seeds occur in many Meliaceae (e.g. *Cedrela, Chukrasia, Khaya, Swietenia*) and Bignoniaceae (e.g. *Markhamia, Tabebuia, Tecoma* and *Spathodea*), and prevail in conifers.

The structure of the wing varies from very thin and membraneous in Bignoniaceae, casuarinas and most pines, to hard and woody in most samaras. Attachment of seed wings to the rest of the seed varies in strength and morphology, and dewinging procedures vary accordingly. The wings of pines are of a type that clasps the seed and is normally lost before germination, usually after wetting. In most other species there is no abscission zone between seed and wing, and most of the wing remains attached to the seed during germination.

Seeds with hairs (floss) occur mainly in Bombacaceae (*Bombax* and *Ceiba* (Kapok)) and Salicaceae (*Salix* and *Populus spp.*). In some *Pterocar-pus* spp. the samara has thin spines.

Wings of conifers are removed by mechanical abrasion during tumbling. Slight wetting by spraying e.g. 1 litre of water to 50 litres of seed during tumbling often facilitates dewinging. Special mechanical dewingers are available where wings are abraded between brushes. Some species like *Abies* spp. have delicate seed that are easily dam-aged during dewinging. These species are preferably dewinged by tumbling in closed drums where the mechanical impact is reduced by slow revolutions where seeds rub against each other. Casuarinas and other species with papery wings may be dewinged by tumbling in cement mixers together with some abrasion material like sand or gravel. The same procedure may be used for removing hairs and spines. If the seeds are mixed with abrading material, it is important that this can easily be removed from the seeds after tumbling and that it does not damage the seed. Abundant hair like that in kapok may be removed by burning.

A very efficient machine for dewinging and detachment of dry appendices such as hairs, floss, arils, floral or fruit parts from the seed is the brushing machine. During operation the seeds are rubbed by revolving brushes against the wall of a cylinder consisting of wire mesh. Rotation speed, distance between brushes and cylinder, type of brushes and mesh wall of the cylinder can be adjusted according to seed type and wing or appendix to be removed.

Large winged seeds like *Swietenia, Entandophragma* or *Triplochiton* spp. are dewinged manually by breaking off the wing by hand or cutting it with secateurs.

Samaras with very hard fruit coats like *Terminalia brownii* and *Pterocarpus* spp. often require a very fierce treatment if wings are to be removed. Some seed processing units have successfully used coffee dehuskers or hammer mills.

CLEANING

After extraction and possible dewinging the seed lot typically consists of seeds mixed with inert matter such as twigs, leaf and fruit fragments, soil particles, empty and foreign seeds, dust, chaff and the like. The aim of seed cleaning is to eliminate all this foreign material to reduce bulk, improve storability and make seeds easier to handle during subsequent processes. The ideal cleaned seed lot consists of all viable seeds of the target species, and is free from any other matter. The degree to which this is achieved is called the purity, usually measured in percentage. 0A purity of 90% means that 90% is seed and 10 % is other seed and inert matter. Some cleaning methods are also used during the process of grading or upgrading in which both germination and uniformity of the seed lot are increased by removing a fraction of seeds of the same species usually the smallest and those with no or poor viability.

Cleaning is a separation process. Some cleaning procedures separate the seed lot into only two fractions, one containing the seed and one containing inert matter to be discharged. Other methods may separate the seed lot into several fractions with various purities. Intermediate fractions typically contain both seeds and inert matter and must be further cleaned. Basically, material can be separated from the seed if it differs in physical characteristics like size, form or gravity. Thus seed cleaning is subject to the trivial precondition that the more the inert matter differs from the seeds in these physical characteristics, the easier it is to separate. And the more similar the impurities are to the seeds, the more difficult they are to eliminate. Variation in seed size and morphology of the seed adds another constraint to seed cleaning: the larger the variation in the seed lot, the more difficult it is to clean. Eliminating inert matter without eliminating viable seeds is difficult for many species. While a purity of say 80% is fairly easy to achieve for most seed, further cleaning can be very hard and laborious. When a certain purity has been achieved, the balance must be considered: either to continue cleaning to achieve a higher purity with the implications of higher processing costs, possible damage to the seeds, and possible loss of viable seeds; or to accept a certain degree of impurity with the implied disadvantages of handling impurities (storage, pre-treatment, sowing etc.), and a possibly reduced price for the seeds.

Seed cleaning typically consists of a series of processes during which impurities are gradually removed and the seed lot concurrently achieves a progressively higher purity. The type, order, and adjustment of the processes depend on seed type and type of impurities. Cleaning typically starts with a pre-cleaning procedure in which large material such as leaves, twigs and entire empty fruits are removed. This may be done in connection with extraction (e.g. tumbling) and is essential for sub-sequent cleaning, especially where machines are employed.

The procedures of manual and mechanical cleaning are based on the same principles of separation according to the physical properties outlined above. Despite the higher direct labour cost of manual cleaning, it is often economical for smaller seed lots since it saves time in adjusting and cleaning the machines after each cleaning process. Cleaning large quantities of seeds is more efficiently done by mechanical equipment, especially where combined machines can be used for extraction and cleaning. In some combined machines the fruits can be fed in one end and the cleaned seed obtained in the other.

Type of cleaning machine must be chosen according to seed type and adjusted appropriately to each seed lot in order to operate efficiently.

If the clean-seed fraction still contains too much debris, or the debris too much seed after the first cleaning, the machine should be adjusted and cleaning repeated. Some seed lots may be efficiently cleaned by one cleaning method e.g. sieving or winnowing. If more than one method must be applied, the order is chosen so that as much debris as possible is removed by the first methods. This is in order to reduce bulk during subsequent processes.

Seed lots which are very difficult to clean to high purity e.g. if they contain a large fraction of empty or insect infested seeds with very similar appearance as healthy seeds, can often be cleaned efficiently by initial grading of the seed lot, usually according to size: once the major portion of debris has been removed by sifting, the seed lot is divided into 2, 3 or more size classes which are then cleaned by one of the other methods. When the individual size classes have been cleaned they are poured together again into one seed lot. This initial grading avoids size differences interfering with other parameters, e.g. specific gravity, and separation consequently becomes much easier in the succeeding cleaning procedures. Often the initial separation makes more complicated procedures like flotation, IDS or PREVAC redundant.

SIEVING

Applicability: Removal of large and small objects from the seed lot. Also used during fruit cleaning (pre-cleaning) and for seed grading. High purity can be achieved for relatively spherical seeds and objects. Less effective for flat or winged seeds.

Physical background: Sieving separates material according to size. Objects may pass an opening larger than their diameter while being retained by an opening of smaller diameter. Asymmetrical objects may pass an opening larger than their smaller diameter when their small diameter faces the opening. Thus an oblong seed will pass an oblong hole, while being retained by a round hole of the same diameter.

Method: The seed lot is sieved through a series of grids with decreasing mesh or hole size. Several types of screens are available. The most common types for small scale cleaning and laboratory use are wire screen grids. Other types, usually used in mechanical cleaners are metal or plastic sheets or wood boards with different hole size and shape. The choice of screen depends on seed type and quantity. Small seed lots of small seeded species like eucalypts are efficiently cleaned using 20 cm diameter laboratory sieves. Larger sieves are used for larger seeds and seed lots. The hole size and shape depend on seed size and shape and type of impurities. In Australia the mesh sizes employed vary from 1 to 4 mm for eucalypts and 3 to 12 mm for acacias.

A screening series consist of at least two sieves:

1. A sieve with openings larger than the seeds typically removes large material like fruit at twig fragments. The holes are adjusted to allowing the largest seeds to pass by their narrowest diameter. Shaking or sliding the seeds over the screens will make them pass.
2. A sieve with openings smaller than the seed retains the seeds while smaller debris passes through. The holes are adjusted to retain the smallest viable seeds.

Sometimes several screens with gradually decreasing mesh or hole sizes may be used and the seeds graded according to size. The grading may be maintained during subsequent cleaning. In some instances small seeds are deliberately discharged.

Many types of mechanical seed cleaners with different and replace-able screens are available. Some smaller laboratory seed cleaners may be supplied with more than 100 screens with different hole size and shape. Large industrial cleaners are normally supplied with a smaller number of screens, but screens can be purchased according to the main species processed. In general round holes are used when the items to be separated differ in width (width is the greater diameter of the cross section of the non-symmetrical seed); oblong holes are used when separation is according to thickness (i.e. the smaller diameter). The appropriate hole size is found by the following method:

1. Place a stack of screens with the correct hole type on top of each other, with the largest opening on top and then decreasing to the smallest opening at the bottom.
2. Pour the seed sample in the upper screen and shake gently to let seed and debris pass holes larger than their diameter.
3. Disassemble the stack of screens and examine the best separation. Choose the appropriate screen size(s).

With small difference in opening sizes more than one of the middle screens may contain both seed and debris. Here one must decide the degree of acceptable contamination. If high purity is required, using several screens (grading) may clean the seed lot. Round or spherical seeds can often be cleaned to high purity by sifting alone, while the method is less effective for flat or winged seeds.

Meshes or holes will inevitably be blocked (blinded) by intermediate size fractions during operation i.e. seeds and particles too large to pass the opening and too small to be left above the screen. The screens must therefore be regularly cleaned, e.g. by brushing. In mechanical clean-ers, screens may be kept clean and blockages of intermediate material prevented during operation by brushing or by placing round rubber balls on the screens. The balls tend to push down or break material get-ting stuck in the holes. A more efficient method is to place the rubber balls on wire-mesh screens with large mesh size under the functional screens. The vibrating movements during operation will make the balls jump up against the screen above and push up material which blocks the holes.

Indented Cylinder Cleaning

Applicability: Separation of seeds and debris according to size or length.

Especially useful for separating twig pieces, pine needles and the like from spherical seeds. Not useful for large seed and seed with large wings or hairs.

Physical background: Sorting according to size is based on the same principle as sieving. Sorting according to length is based on the difference in gravity point of short and long seeds. When indented in a short pocket in horizontal position, short seeds have their gravity point within the pocket and remain there, while long seeds with gravity point outside the dent fall out.

Method: The indented cylinder consists of a cylinder with numerous indentions in its inner surface, revolving round a sloping axle. Above the axle along its length is a fixed sloping trough. During operation seed to be sorted is fed in at the upper end of the cylinder and slowly moves downward to the lower end. As the cylinder revolves slowly, seeds fitting into the pockets are carried upwards. Separation according to size and length is illustrated in the inset.

When used for size separation, a cylinder with pocket size that will allow only small size seeds to fit into the pockets is used. Large seeds, which do not fit into the pockets, remain in the cylinder and move downward to the lower end. Small seeds are carried upwards and fall into the trough.

When used for separation according to length, both short and long objects initially fit into the pockets and are carried upwards. However, long objects like twigs or straw, fall out before they reach the trough and thus remain in the cylinder.

Cylinders with different pocket sizes are available and are easy to change according to seed size. The flow through the cleaner is adjusted by adjusting the slope of the cylinder.

Winnowing and Blowing

Applicability: Removal of dust, chaff, wings or other light material from seed lots. Less effective for light winged seed.

Physical background: Blowing and winnowing separate material according to different gravity (weight/volume) and surface/volume ratio. An object with a large gravity (e.g. a stone) or a small surface/volume ratio (e.g. spherical) will fall faster vertically and be moved a shorter distance by a horizontal air current than an object with a smaller gravity (e.g. wood) or large surface/volume ratio (e.g. a leaf).

Method: In its simplest form of winnowing, seeds are held in large flat baskets. The seeds to be cleaned are thrown up into the air, and the wind will then blow away light matter like dust, wings and leaf fragments, while the heavier seeds will fall back into the basket. This is the traditional way of cleaning grain and can be quite effective for small seed lots. Natural wind displacement can also be used by slowly pouring the un-purified seeds from a certain height into a pile, allowing the wind to blow away light matter while the seeds fall. Since natural wind velocity varies, the method is not always efficient.

In the mechanical form winnowing works with an artificial and normally adjustable air stream. Table fans or vacuum cleaners can create simple artificial air currents. More sophisticated cleaners use large stationary fans or propellers. The strength of the air current is adjusted according to seed type and debris. To work effectively, the seed must be fully exposed to the air current. There are two variants. Either the seeds pass vertically through the air current or shaking them on an undulating surface over which the air current runs elevates the seeds. Normally the grids on which the seeds are shaken are provided with holes, thus combining winnowing and sifting. Winnowing sorts seed into a gradient with heavy particles (seed) closest to the air source and the lightest farthest away. The intermediate fraction contains a mixture of light seed and debris with decreasing purity away from the air source. This fraction is sometimes sorted by cleaning machines into several sub-fractions that are subsequently re-cleaned by winnowing or other methods.

A simple winnowing chamber, primarily designed for cleaning seeds of Swietenia macrophylla but with wide applicability. The chamber consists of 1) a wind funnel from where an air current is created by a fan, 2) a central section with a formica top sloping against the air current, and 3) a gauze cage where light debris is collected. The uncleaned seeds are fed from the top above the central slope and pass through the air current. Light matter is blown into the gauze cage, the seeds fall down on the slope and roll or slide down. Slope, height of drop and strength of air current can be adjusted according to seed type.

The physical principle of air current displacement is also used in seed blowers, which are mainly used for small seed types such as eucalypts and casuarinas. The seed lot to be cleaned is placed in a vertical cylinder connected to an electrical powered air current at the bottom. The upwards air current will displace all light material like chaff and wings to the top while the heavier seed are collected at the bottom. The cylinder can be emptied in sections so that the mid section, which typically contains a mixture of seed and debris, can be re-cleaned.

Specific-gravity Separators

Applicability: Separation of seeds and objects differing in weight, density and surface characteristic. High precision with consequent high purity can be achieved for many species.

Physical background: The separation is, as for preceding procedure, based on variation in specific gravity and surface/volume ratio. In specific gravity separators the seeds and impurities are exposed to forces greater than gravity. The balance between gravity and a force in the opposite direction will stratify the mixture.

Equipment and method: Two variations of specific gravity separators are available, the oscillating table and the pneumatic table separator.

1. *Oscillating table:* The separator consists of a slightly inclined table with zigzag partitions along its length. During operation the table is shaken sideways so that objects placed on the table will be struck by the partitions. Seeds placed in the middle of the table will tend to move downwards by gravity but the strikes by the partitions will tend to move them upwards. Separation of the seeds is based upon the balance between these two forces. For light seeds the striking impact of the partitions overcome gravity and hence move them upwards. For heavy seeds the striking impact is insufficient to overcome gravity and the seeds slide or roll downward. The sloping of the table and the sideways movement can be adjusted for different seed type. As the seeds move on the deck during operation they are also influenced by their surface structure. A surface with a high friction is slowed down during movement and follows the stream of heavy seeds.
2. *Pneumatic table separator or specific gravity table:* The separator consists of a slightly inclined table with a porous surface, e.g. a woven linen cloth, connected to a compressed air source. During operation an air stream is forced up through the table making the seed float on an 'air cushion'. The seed mix will tend to stratify vertically on the 'air cushion' with the heavy seeds at the bottom. Vibrating or oscillating movements of the inclined table inflicts separation. Since the heavy seeds are at the bottom, they will be hit by the surface of the table, which will tend to

move them upwards. Hence, light particles will float over the edge of the table at the lower end whereas the heavy particles fall off the table at the upper end.

Pressure of the air stream, inclination of the table (in two directions), and surface of the deck can be adjusted to seed type. Size of the deck varies in different models, the smaller with maximum length of some 25 cm, the larger models about 125 cm.

Friction Cleaning

Applicability: Separation of seeds that differ in form or surface structure, e.g. round seeds to be separated from flat seeds, or smooth seeds to be separated from rough seeds.

Physical background: Shape and surface structure make material slide differently down a slope with high friction material. An object with a high surface friction (e.g. a leaf) can stay on a steep slope while an object with low friction (e.g. smooth paper) will slide down. Further, an object with a low gravity point may stay on the same slope while an object with a high gravity point will roll down. Since spherical objects have both a small friction and a high gravity point, they will roll down a slope with a relatively small angle.

Method: Friction cleaning in its simplest form is carried out by letting the seeds move on a sloping cloth frame. Flat objects will remain on the cloth while round objects roll down and are collected at the bottom. In a mechanical derivation of this system, seeds are fed onto a sloping rotating cloth. Round seeds roll down the cloth and are collected in one fraction, flat seeds are carried up the slope into another fraction. Feeding of the seed mixture to the separation cloth must be sufficiently slow, so that the seeds can be carried away or roll freely. The rotation speed must be adjusted so that the seeds move smoothly without jumping. The slope is adjusted so that the most effective separation is achieved.

Where the difference is only in surface texture and not in gravity point, a vibrator separator is used. The vibrator separates seeds and other matter according to their surface characteristics on a small deck with a rough surface. A rough surface of the seed will tend to grip the surface of the sloping deck and move the seed upwards when the deck vibrates, while a seed with a smoother surface slides down. Seeds will thus tend to stratify according to their surface characteristic at the outlet end of the table. The vibrator separator has several options of adjustments which can make separation very accurate viz. 1) the speed of deck vibration, 2) the roughness of the deck (different decks are available with surface ranging from linen to rough sand paper), 3) degree of side and end tilts, 4) the rate at which the seed is fed onto the deck, and 5) the arrangement of the outlet gates.

Flotation

Applicability: Separation of mature viable seeds from chaff, empty seed, filled dead seed, immature or insect-damaged seed or other particles with a specific gravity smaller than that of seeds. While specific gravity differences in air (winnowing, specific gravity separators) may be obscured by surface and volume differences, flotation is independent of the latter. The flotation technique is also applicable if gravity difference between sound seed and inert matter is very small.

Physical background: An object placed in a liquid medium will float if its specific gravity or density (weight/volume) is smaller than that of the liquid, and sink if it has a higher specific gravity than the liquid. Most seeds have a density slightly below 1.0 when dry, while their gravity is slightly above 1.0 when imbibed. That implies that they will tend to float in water (specific gravity = 1) when dry and sink when imbibed. The IDS and PREVAC methods below use the fact that sound seed absorbs and desorbs water at a different rate than dead filled seed and damaged seed.

Method: Two methods are described by Simak, viz. density method and absorption method.

1. *Density method:* A liquid with a density or specific gravity between those of full and empty seeds (usually below 1.0) is used. Full seeds will sink and empty seeds and light debris float. To work effectively the specific gravity of the flotation medium must be adjusted to a level between the light and heavy material. Table 3 lists a number of liquids that have been used for separation by liquid flotation. A mixture of two liquids with different densities, e.g. n-pentane and ethanol, makes a solution with intermediate density; adding a soluble compound to a liquid (e.g. salt to water) normally increases its density. It is thus possible to adjust the density of the liquid quite precisely for optimal separation. In separation of filled and unfilled seed of Araucaria cunninghamii a mixture of 95% ethanol and n-pentane in different proportions were used. It was shown that the density of filled and unfilled seed varied with clones, and the density of the flotation medium thus had to be adjusted accordingly.

   Some flotation media may affect seed viability. It was shown by Simak that while absolute alcohol had no negative effect on germination, lower concentrations could apparently damage the seed. Barnett showed that ethanol had a negative effect on storability of some pine species. Short and long term effects on viability of other organic flotation media have been documented by Hodgson. The liquid used for separation must obviously be harmless to the seeds. Alcohols are poisonous to seed embryos, but harmless as long as it is only in contact with fruit or seed-coats. Therefore, potential damage is dependent on the liquid itself plus the absorption rate, which in turn depends on the seed-coat structure and the period of exposure. For example, a hard coated species may tolerate short exposure to a poisonous medium, while a longer exposure or a quicker absorption may be detrimental.

Table 3. Specific gravity of some liquids used for separation by flotation.

Medium	Specific gravity
Pure water	1.0
Abs. alcohol (ethanol)	0.791
95% alcohol (ethanol)	0.806
50% ethanol	0.90
Diethyl ether	0.714
Petroleum ether	0.657
n-penthane	0.626
Mixture 95% ethanol and n-pentane, 3:1	0.76
Mixture 95% ethanol and n-pentane, 12:13	0.71
Linseed oil	0.93

2. Absorption method. Seeds to be sorted are poured into water (specific gravity 1.0). The seeds will initially float, but mature viable seeds will absorb water and sink after some time, from a few minutes to several hours. Empty, immature or damaged seeds and other light material may remain floating and can be skimmed off after an appropriate period of time. Often, however, empty and damaged seeds absorb water at the same rate as sound seeds and will sink accordingly. In this case the seeds are re-dried shortly (a few minutes to several hours) during which period the empty and damaged seeds will lose water more rapidly than healthy seeds. During a second flotation healthy seeds sink while immature or damaged seeds float. The mature sound seeds must be re-dried after separation if the seeds are to be stored before sowing.

The absorption method has been further developed into two practical methods, the IDS and PREVAC methods. The methods have been developed and used for pines but may be applicable to other species as well.

IDS (Incubation-Drying-Separation). The method is mainly used as a pre-sowing separation to eliminate seeds that have lost viability during storage. In the IDS method the seeds to be separated are initially soaked to allow imbibition, then incubated at optimal germination conditions (e.g. light, optimal temperature and high humidity). After 2-3 days of incubation the seeds are partially dried, and separated by flotation. Seeds with low viability tend to lose water faster during drying than full viable seeds; the seed lot can thus be sorted according to probable seed viability.

The method has one main limitation viz. that unintentional germination can be difficult to avoid during incubation. This is likely to affect storability and the method should therefore not be used before storage. Germinating seeds are also prone to damage by over-drying after incubation.

PREVAC (Pressure-vacuum). A method used to separate seeds with mechanical damage from sound seeds. Dry (unimbibed) seeds are exposed to low pressure (vacuum) for 1 to 20 minutes while lying in water. When the pressure is released, mechanically damaged seed (e.g. with cracks or part of the seed-coat missing) absorb water more quickly that undamaged seed. During subsequent flotation, damaged seeds tend to sink while undamaged seeds tend to float.

In Australia a method using a combination of the density method and the absorption method is used for the separation of live seeds of Eucalyptus pilularis from chaff and dead seeds. The seeds are initially pre-imbibed in water for 2-4 days, then separated in a sugar solution.

SEED GRADING

While the objective of seed cleaning is to improve purity by eliminating non-seed material and foreign seed from the seed lot, the purpose of grading is to improve the average physiological quality of the seed lot by removing seed of the same species with low quality. Such seed may be empty seed, immature seed, damaged or dead seed or seed developed after self-fertilisation. In the latter case the removal also serves to improve the genetic quality of the seed lot. Sometimes a larger fraction of small yet viable seed is deliberately removed from the seed lot based on an assumed correlation between seed size and vigour.

A positive correlation between seed size and seedling size/vigour has been documented for several species. Often large seeds tend to germinate faster and produce larger and more vigorous seedlings than small seeds of the same species. The difference is usually pronounced during the early stages of seedling development but tends to become smaller and may disappear after one or more growth seasons. From a silvicultural point of view, swift and uniform germination and seedling establishment is obviously an immediate advantage, albeit not necessarily correlated with ultimate yield.

Grading according to size can be useful to assure a more uniform germination speed and seedling growth within each grading class. A uniform seed size facilitates sowing with sowing machines and a uniform germination and seedling growth rate will imply fewer cullings. Grading, with elimination of the lightest seeds, may influence the genetic composition of the seed lot by selectively removing seeds of families with relatively small seeds.

Seed grading may in practice be an extension of the seed cleaning process because the small and light seeds are removed together with chaff and other impurities. Methods of grading must, however, be adjusted much more precisely since the physical difference of seeds within a species is likely to be much less than between seeds of different species or seeds and extraneous matter.

ADJUSTING MOISTURE CONTENT FOR STORAGE

After having obtained clean pure seed the final processing procedure is to adjust moisture content for seed that is to be stored for any length of time. The appropriate moisture content varies with species and potential storage period. Orthodox seed, which includes most dry-zone species plus most humid zone pioneers, has no lower moisture content tolerance level. A moisture content of 6-8% normally assures safe storage; lower moisture content is only appropriate for long term storage at very low temperature. Recalcitrant seed, which prevails among humid zone climax forest species, does not tolerate much desiccation and must be stored with a high moisture content for the shortest possible time. A large intermediate group (semi-orthodox, semi-recalcitrant or 'intermediate' seeds) shows varying levels of tolerance. While orthodox seed is either further dried or maintained at the moisture level achieved during processing, recalcitrant and intermediate seed may need re-moistening for safe storage. The moisture content for these species is adjusted to the minimum for safe storage. The actual moisture content of the seed lot is measured before storage either by a calibrated moisture meter or by the ISTA oven method.

Orthodox Seed

Most orthodox seed can be stored safely for at least 1-2 years at a moisture content of 8-10% or below. Potential storage period is prolonged by cold storage. For long term storage at sub-zero temperatures a moisture content of 2-4% is desirable. Drying may be carried out by natural sun drying or artificial heating with a dry air current, which is usually supplied by electrical appliances. The relation between moisture content and heat should be recalled: moist seed is less tolerant to heat than dry seed. If the initial moisture content is high, e.g. after moist extraction or cleaning, care should be observed not to overheat seed during the initial drying stage. The following drying temperatures may be suggested:

1. Air temperature 30-35°C until moisture content has lowered to 10-12%.
2. Air temperature 35-45°C until moisture content has lowered to 5-10%.
3. Air temperature 45-55°C until moisture content has lowered to 3-5%.

Once the seeds have been dried to an appropriate storage moisture content, they should be stored in air-tight containers as soon as possible to avoid re-gain of moisture from the air. The rate of re-absorption depends on species. In Nepal, Napier and Robbins report that Alnus nepalensis under humid conditions may re-absorb moisture within hours after drying.

Recalcitrant and Intermediate Seed

Although recalcitrant seed has high water content when shed, most species undergo some kind of maturation drying. The lowest safe moisture content varies with species and specific information should be consulted. For some recalcitrant seeds the rate of moisture reduction has shown influence on subsequent storability. Triplochiton scleroxylon, whose seed normally lose viability in a matter of weeks, was shown to tolerate drying down to 8% m.c. provided the moisture content was lowered by about 1% per hour. The seeds could then be stored at low temperature and maintained viability for more than a year. Thailand provenances of neem (*Azadirachta indica*) maintained a high viability (~ 60%) after 6 weeks of storage when the seeds were sundried for 2-3 days. Both shorter (1 day) and longer (7 days) duration of drying gave inferior germination. No major differences in storability were found after 8 weeks' storage following alternative drying methods and rates. Thompset found, however, that the rate of drying was of no importance for the storage behaviour of *Araucaria hunsteinii*. The effect of drying rate on most other recalcitrant seeds still needs to be revealed.

Sensitive seeds may suffer from accidental desiccation damage caused by processing. Such damage cannot be restored by re-moistening, but desiccation may be minimised by storing the seeds with a safe moisture content and in an environment where further desiccation is avoided. Re-moistening is technically the opposite of drying. For recalcitrant seed it is, however, complicated because a balance must be maintained where moisture content increases to the desired level without initiating imbibition and germination. Re-moistening by absorption from humid air may be easier to control than submerging the seeds in water. Regarding safe moisture content for various species.

POTENTIAL SEED DAMAGE DURING PROCESSING

Seed processing aims to achieve a balance between maximising effectivity (extraction, cleaning, protection against deterioration) and damage to the seeds. In practice, processing always implies a risk of damage or injury to some seeds. Damage may occur in various ways:

1. *Mechanical damage:* Usually on the seed-coats but occasionally on the embryos with well developed seed cotyledons. Generally, spherical seeds and small seeds tend to suffer less damage than elongated or irregularly shaped seeds.
2. *Heat damage:* Often occurring by exposure to high kiln temperature for

extracting seeds from cones, or deliberate burning for removal of fruit or seed hairs. Fatal high temperature can also occur during fermentation of fruit pulp. Moist seeds are more prone to heat damage than dry seeds, and recalcitrant seeds are, accordingly, sensitive to heat damage.

3. *Chemical damage:* Sometimes occurring during separation by flotation in organic liquids. Other potential sources are fungicides.
4. *Water:* Prolonged submersion in water, e.g. to soften the fruit pulp may hamper respiration of the seeds. Prolonged soaking may also cause imbibition and initiate germination in seeds with no dormancy.

The severity of the damage depends on extraction / handling procedure and of seed type:

1. The more fragile the seed, the more sensitive it is to damage. Seeds with thin seed-coats or large cotyledons without or with little enclosing endosperm are easily damaged by some processing methods.
2. The more frantic the process, the higher the potential damage. Threshing and beating e.g. of indehiscent pods imply a potential risk of breaking the embryo. Especially sensitive is the attachment site of the cotyledons to the embryonic axis. Mild impact to seed-coats can have a beneficial influence on germination by breaking physical dormancy.

Studies on the effect of seed processing on seed quality have mainly concentrated on conifers. In comparison with other species, conifers require quite a lot of handling and processing in order to extract the seeds, and their seeds are often subsequently dewinged to ease handling. At the same time seeds of many conifers are fragile and easily damaged by handling. Handling damage while the seeds were still enclosed in the cones has been reported for *Abies* spp. Throwing sacks of cones to the ground during collection was sufficient to cause quality reduction.

Damage during dewinging may occur when seeds are tumbled e.g. together with debris. Heavy damage has been encountered in seeds of *Abies lasiocarpa* where up to 50% of the seeds was lost. Excessive dewinging resulted in dull dusty seed-coats, susceptible to mould and resulting in weak seedlings.

Temperature damage of conifer seeds during extraction in kilns has been widely reported. Most sensitive are immature and moist seeds, and temperature damage is accordingly most likely to occur during the early phases. Potential damage also depends on length of exposure.

Wang *et al.* found a substantial loss of viability of *Pinus contorta* var. latifolia seeds by exposing the cones to a scorching temperature of 220°C for more than 1.5 min. Shorter exposure did not impair viability, possibly because the temperature inside the seed did not rise to a fatal level for the embryo.

A few investigations suggest that seed damage may occur during chemical treatment. Fumigation with carbon disulphide or hydro-cyanic acid for killing *Megastigmus* spp. in conifers affected viability. Some alcohols used for flotation had a negative effect on the viability of *Pinus* seeds; others caused no loss in viability.

There is little documentation on mechanical damage to forest seed during processing, but some parallels may be drawn from experience from agricultural seeds. In maize (*Zea mays* L.) various degrees of damage on seed-coat, endosperm and embryo could be detected in 89% of the seeds after processing. The injuries affected germination, seedling development, susceptibility to diseases, plant growth and development and grain yield. The germination rate and seedling quality was influenced by the location of damage, the embryo and especially the central part being the more sensitive.

Minor damage to seeds during processing may not immediately affect viability but may cause reduced seedling vigour and misshapen seedlings. Damage also affects storage potential since injured or deeply bruised areas may serve as centres for infection. This may partly be caused by an accelerated progressive deterioration or interaction with other deteriorating factors, e.g. increased susceptibility to fungal infection through cracks in the seed-coat. Injuries to or near delicate parts of the embryo are prone to both primary and secondary deterioration.

While heat damage is most likely to affect moist seed, dry seed seems more susceptible to mechanical damage. Therefore it must be advised that seeds should only be moderately dried before mechanical treatment e.g. extraction, dewinging and cleaning.

SAFETY PRECAUTIONS DURING PROCESSING

Like seed collection, processing implies both general and specific safety hazards. Processing staff should be familiar with these potential risks and observe appropriate precautions.

1. *Fire danger:* Dry fruit parts, resin and dust released during processing of dry fruits can easily catch fire and therefore pose a fire hazard. Use of artificial heat or other electric appliances during extraction increases the danger. Dust may catch fire when coming into direct contact with glow wires or the like. Therefore, heat sources should be safely shielded and dust removed regularly during processing. Water and/or fire extinguishers should be readily available at the seed-processing unit.
2. *Respiratory, eye and skin irritations:* During processing, floral parts, fungal spores, dry pulp and other fine particles become suspended in the air and form what is commonly known as dust. Some species, e.g. *acacias*, are known to release especially large amount of dust when threshed.

Because dust is dry, it causes a general irritation of eyes, nose, and skin with resulting itchiness, coughing and sneezing. For most people this is merely annoying, but for some people some dust elements cause allergic reactions. Dust problems can be minimised by appropriate ventilation, possibly by outdoor handling. Extractors should be placed as close to the sources as possible, e.g. near threshing machines. Staff working with species or equipment with particular dust problems should be provided with dust masks and possibly also dust glasses.

Softening the fruit flesh of sugar palm (*Arenga pinnata*) by soaking prior to depulping requires caution. Decomposing fruits develop a fluid causing intense itching and burning whenever it comes into contact with the skin. Also contact with the seed-coat can cause skin irritation. Processing of seeds of *Platanus* spp. and several species of the family Boraginaceae are known to give similar skin irritations. Rubber gloves must be used when handling these fruits.

3. *Mechanical equipment:* The risk of accidents with mechanical equipment such as threshers and grinders can be greatly reduced by safe construction and maintenance of the equipment and appropriate training and instruction of the operators. Potentially dangerous mechanical or electrical parts (rotating devices, cords etc.) should be shielded with screens. Screens should be mounted in the front of inlets to e.g. threshers and operators should observe a safe distance. Emergency switches should be placed near the place of operation so that machines can easily be stopped in case of an accident.
4. *Poisonous fruit pulp:* Some fruits like *Strycnus* spp. have poisonous pulp, fatal to humans and livestock. Removed pulp and water used for extraction must be discharged and disposed of safely.

During processing the fruits and later the seeds pass through a number of processes, they are unloaded and loaded into different containers and processing equipment, and often handled by a number of people. The risk of losing or accidentally mixing labels are obviously high, especially when handling a number of minor samples of the same species, e.g. single tree collections or provenance collections. A system must be created to minimise the risk of losing seed identity. Handling of labels is, in many cases, as important as handling of the seed itself. Simple routine procedures are recommended. If some members of the staff are not able to read the labels, they should still be able to maintain the routines. Some points are summarised below:

1. Two labels should always follow the seed lot during collection. One is placed outside the container, one is put inside together with the seeds. The labels should be written with water-repellent ink, the labels should be resistant to some degree of moisture.

2. Labels that are no longer valid should be discarded to avoid later confusion, e.g. if new labels are written because the old ones become difficult to read, or if several seed lots are mixed.
3. When fruits or seeds are poured into e.g. trays, depulping or cleaning machines where the label cannot be kept with the fruits or seeds, or where it would be easily lost by wetting or blowing away, the labels should be clipped or stuck to the processing equipment. Once the particular processing part has been concluded, the label is replaced together with the processed seeds.
4. Partly processed seeds are preferably put into the same containers again. After reduction of the major bulk (e.g. after extraction) fewer, smaller or different types of container may be used. The new containers must be labelled, and redundant labels discharged.
5. If part of the seed is fully processed and another part needs additional processing, the two parts must be separated and labelled individually, e.g. A, B, C,...
6. Discharged labels should be torn or removed completely from the processing site (not just thrown on the floor) in order that they will not later be confused with valid labels.

A second point in maintaining identity relates to the risk of physically contaminating the seed lots. If the seeds are to be used for trials, contamination may completely distort the results. It is rarely possible to clean a seed lot for seeds of the same species, and separation of seeds of some species with very similar seeds may also be impossible. Therefore, contamination is often irreversible.

The chances of contamination during seed processing are many. Light seeds may blow from one seed lot to another; perforations in containers or trays may cause seeds to slip from one container or tray to the next if stacked; seed may be stuck in storage containers or processing equipment, especially tiny seeds like eucalypts, Anthocephalus or casuarinas. Hygiene routines must be followed:

1. The same containers are used before and after part-processing.
2. Emptied containers are thoroughly cleaned before they are used for any other seed lot. Bags are turned inside out to be cleaned in stitching and corners.
3. Processing equipment is thoroughly cleaned after each process. Brushing, the use of compressed air or a strong water current is often necessary for appropriate cleaning.

REFERENCES

Barnett, J.P. 2021. "Flotation in ethanol reduced storability of Southern pine seeds". *Forest Science,* 17, 50-54.

Chaisurisri, K., Ponoy, B., Wasuwanich, P. 2016. Storage of *Azadirachta indica* A. Juss. seeds. The Embryon, Vol.2, No. 1, 19-27.

Doran, J.C., Turnbull, J.W., Boland, D.J, Gunn, B.V. 2013. *Handbook on seeds of dry zone acacias. Guide for collecting, extracting, cleaning and storing the seed and for treatment to promote germination of dry-zone acacias.* FAO, Rome

For. Com. Tasmania 2014. *Seed manual. Procedures for seed collection, handling, storage, purchase and accounting*. Forestry Commission, Hobart, Tasmania.

Granhof, J. 2014. *Seed extraction 2. Extraction of pine seeds by means of sun drying on elevated trays followed by tumbling.* DFSC Technical Note No. 17. Danida Forest Seed Centre. Denmark.

Kabay, D. and Lewis, A. 2019. "Collection, handling and storage of Australian native plant seed". In: *Germination of Australian native plants seed.* (Langkamp, P.J., ed.). Inkata Press. pp. 20-30.

Napier, I.A. and Robbins, A.M.J. 2018. *Seed storage in the nursery* (*Prunus, Alnus* and *Pinus* spp.). Banko Janakari, Vol. 1: 1, 13-20.

Stubsgaard, F. and Moestrup, S. 2021. *Seed Processing*. Lecture Note No. C-7. Danida Forest Seed Centre, Denmark.

Thomson, J.R. 2019. *An introduction to seed technology*. Leonard Hill.

Tompsett, P.B. 2022. "The effect of desiccation on the longevity of *Araucaria hunsteinii* seed". *Annals of Botany* 50, 693-704.

Veira, C.P., Veira, R.D. and Paschoalick, J.H.N. 2014. "Effects of mechanical damage during soybean seed processing on physiological seed quality and storage potential". *Seed Sci. and Technol.* 22: 3, 581-589.

11

Seed Quality

Seed is a living product that must be grown, harvested and processed correctly to maximize its viability and subsequent crop productivity. For the yield potential of any rice variety to be realized, good quality seed must be sown. Good quality seed can increase yields by 5-20%. The extent of this increase is directly proportional to the quality of seed that is being sown. Seed quality can be considered as the summation of all factors that contribute to seed performance.

High quality seed enables farmers to attain crops, which have:

1. the most economical planting rate
2. a higher percentage of seeds
3. emerging in the field a minimum of replanting
4. a vigorous seedling establishment
5. a more uniform plant stand faster growth rate, and
6. greater resistance to stress and diseases and
7. uniformity in ripening.

FACTORS AFFECTING SEED QUALITY

Seed quality is determined by a number of genetic and physiological characteristics. The genetic component involves differences between two or more genetic lines, while differences between seed lots of a single genetic line comprise the physiological component.

The genetic factors that can influence quality include:

1. genetic make-up
2. seed size and
3. bulk density

The physical or environmental characteristics include:injury during planting and establishment

1. growing conditions during seed development
2. nutrition of the mother plant
3. physical damage during production or storage by either machine or pest ,
4. moisture and temperature during storage and
5. age or maturity of seed

Deterioration in seed quality may begin at any point in the plant's development stage from fertilization onward. Seed quality depends upon the physical conditions that the mother plant is exposed during its growth stages, as well as harvesting, processing, storage and planting. Temperature, nutrients and other environmental factors also affect seed development and influence seed quality.

High quality seeds are the result of good production practices, which include:

1. proper maintenance of genetic purity
2. good growing conditions
3. proper timing and methods of harvesting
4. appropriate processing during threshing, cleaning and drying
5. appropriate seed storage and seed distribution systems

CHARACTERISTICS OF GOOD SEED

Seeds of high quality should be true to its kind or variety, contain a minimum of impurities and have high establishment rates in the field. The main criteria for describing seed quality can be considered under the following:

1. Varietal characteristics
2. Seed lot characteristics
3. Seed viability

Varietal Purity Characteristics

Varietal purity refers to the genetic or cultivar purity and can be described by its physical, chemical and crop attributes.

Seed Factors

The true seeds of a variety have are characterized by specific traits such as the length and width, shape, size, color and aroma. Laboratory tests to quantify these characteristics include physical measurements of the length and width, 1000 grain weight, color measurements using the naked eye and colorimeter and smell tests both ocular and machine.

Chemical Analysis

Various chemical tests can also be used to distinguish between varieties. Factors such as amylase content, alkali digestion color, gel consistency and brown rice protein help distinguish between varieties.

Crop Factors

The crop characteristics that are normally consistent include plant height, time to maturity, plant color and plant growth habit. Slight variations may occur especially in harsher environmental growing conditions.

Seed lot Characteristics

A description of the seed lot includes the level of impurities, seed size, and damaged, deformed or diseased seed.

'Impurities' is described by the degree of contamination through weed seeds, seeds of other crops or species, and inert material such as stones, dirt or twigs. Impurity is expressed on a percentage basis by weight.

'Seed size', plumpness and/or fullness, are generally desirable seed characteristics. They indicate that the seed has the potential to produce vigorous seedlings under favorable conditions.

'Damaged, deformed or infected seed'. High quality seed should be free from seed-borne diseases, insects and other extraneous matter. They should also be free from various types of mechanical injury that reduce germination and seedling vigor.

'Red Rice' kernels are not acceptable in the rice market place. Kernels should not have more than 25% of its surface area colored red or red streaked.

'Discoloration' can be caused by fungi, bacteria or environmental conditions such as high humidity and temperature (e.g., 27-35°C). There are two types of discoloration on rice seeds, 1) spotting caused by certain diseases (brown spot) or insects (rice bug feeding damage) and, 2) discoloration on a portion or portions of rice seeds caused by Curvularia (black kernel) or sometimes by Alternaria padwickii (stackburn disease) or by sheath rot on certain occasions. Other rice diseases do not produce consistent discoloration. We consider any seed having more than 0.5% discolored or spotted seed surface as discolored. Discoloration can occur on fully or partially filled or empty grains. Laboratory tests are required to confirm whether the cause is fungal or bacterial. If the grain is discolored after milling, then the problem was most likely environmental.

Seed Viability

The viability of the seed in the field will be determined to a large degree from its stored moisture level, Germination potential and its vigor.

'Moisture content' has a marked influence on the life and vigor of the seed. Moisture content should be less than 14% and preferably less than 12% for extended storage times.

'Germination percentage' expresses the proportion of the total number of seeds that are alive. It is determined through controlled tests and actual counts of the number of seeds that germinate. Many varieties have a dormancy period immediately after harvest. Stored under traditional open systems the germination rate of most rice seed begins to deteriorate rapidly after 6 months.

'Seed vigor' provides a very good estimate of the potential field performance, and subsequently, the field planting value. While the speed of germination varies slightly across varieties, seeds should imbibe (absorb) moisture and within 2 days produce a root (radicle) and the first leaf (plumule) within 4 days. At this point the seed is considered to have germinated. The ability of the germinating seed to continue to grow and survive then determines crop establishment. Seed vigor has been defined as the sum total of those properties of the seed, which determine the level of activity and performance of the seed during germination and seedling emergence. Seed vigor is an important factor that often results in poor seedling establishment. Seeds low in vigor generally produce weak seedlings that are susceptible to environmental stresses. Whereas, a high level of vigor in seeds can be expected to provide for early and uniform stands which give the growing seedlings the competitive advantage against various environmental stresses.

SEED CERTIFICATION

The purpose of seed certification is to maintain and make available to farmers, high quality and genetically pure seeds of superior cultivars. Only those cultivars that are of superior genetic makeup, multiplied to maintain purity and identity, are normally eligible for government certification. Certified seed is high in genetic purity, high in germination and vigor, and of good quality (i.e., free from disease and from damaged or immature seeds).

Cultivar purity is the first consideration in seed certification, but other factors such as the absence of weeds and diseases, viability, mechanical purity, and grade are also important. One of the most effective ways to limit the distribution of weeds is to plant weed-free seed. Planting disease-free seed can reduce losses in the same way. Properly, cleaned and graded seed is easier to plant and gives more uniform stand. Thus, seed certification has been designed not only to maintain genetic purity of superior cultivars, but also to establish and maintain reasonable standards of seed condition and quality.

Seed Testing—Seed samples are collected and submitted for laboratory analysis after drying and processing.

Tests conducted include:

1. Varietal purity,
2. Weed and other crop seed,
3. Inert material
4. Other varieties
5. Red rice and
6. Germination.
7. Moisture content

For seed certification, field inspectors must check:

1. Crop land requirements (e.g., previous crop and isolation distances to ensure no pollen mixes and or volunteers from a different previously planted cultivar)
2. The source of the crop's seed (to ensure cultivar purity)
3. Sowing dates, pattern, spacing, and rouging (to ensure crop kept clean and free from contamination.)
4. purity of cultivar (i.e., no cultivar mixes)
5. Harvesting methods (to prevent mixtures)
6. Postharvest operations to ensure correct cleaning, drying, sampling, tagging, labeling and sealing

Classes of Seeds

'Breeder Seed'—this is the seed of a new variety that has the highest purity, produced and developed, controlled and provided directly by the breeder or his/her institution for further multiplication.

'Foundation Seed'—this is the progeny of the breeder seed, produced by trained officers of an agricultural station in conformity with regulated national standards and so handled to maintain genetic purity and identity of the variety.

'Registered Seed'—this is the progeny of the foundation seed grown by selected farmers, so handled to maintain genetic purity and identity and has undergone field and seed inspections to ensure conformity with standards.

'Certified Seed'—this is the progeny of foundation, registered or certified seeds, so handled to maintain sufficient varietal identity and purity, grown by selected farmers under prescribed conditions of culture and isolation and subjected to field and seed inspections prior to approval by the certifying agency. Harvest from this class is used for commercial planting.

Classes of Seeds: Examples of Standards

Different countries set standards for the different factors considered in certifying different classes of rice seeds.

Table 1: Official Standards for Seed Certification in Arkansas

Factor	Breeder	Foundation	Registered	Certified
Pure seed (%)		98	98	98
Other varieties (grains/500g)		None	None	2
Other crop seeds (max/500g))		None	None	2
All noxious weeds		None	None	None
Total weed seed (%)		0.03	0.03	0.08
Inert matter (max %)		2	2	2
Red rice (grains/500 g)		None	None	None
Germination (% minimum)		80	80	80
Moisture content (max %)		14	14	14

Table 2: Official Standards for Seed Certification in Philippines

Factor	Breeder	Foundation	Registered	Certified
Pure seed (%)	98	98	98	97
Other varieties (grains/500g)	0	2	5	10
Weed & other crop seed (%)	0	0	0.05	0.1
Inert matter (%)	2	2	2	3
Red rice (grains/500 g)	0	0	1	2
Germination (% minimum)	80	80	80	80
Moisture content (%)	14	14	14	14

Table 3: Standards for seed certification in Myanmar

Factor	Breeder	Foundation	Registered	Certified
Pure seed (%)	99	98	98	97
Weed seeds (no's/500gm)	3	5	10	10
Red rice (grains/500 g)	0	1	3	5
Germination (% minimum)	90	90	85	80
Moisture content (%)	13	13	13	13

SEED PRODUCTION: FIELD MANAGEMENT

Planting Stock

In most varieties, *'breeder seed'* must be planted to produce foundation seed. *Foundation seed* must be planted to produce registered seed and registered seed must be planted to produce certified seed. In varieties where there is no registered class, foundation seed must be planted to produce certified seed.

Field Inspection

In most countries when a crop is intended for certification it must be inspected prior to harvest. A representative of the certifying agency inspects the crop at least once during the vegetative stage and normally once at pre-harvest stage. The crop may be refused certification due to unsatisfactory appearance caused by weeds, poor growth, poor stand, disease, insect damage and any other condition, which prevents accurate inspection or creates doubt as to the identity of the variety. Each country has field standards for the allowable number of off types, unacceptable weed types and numbers of diseased plants. The field standard for Myanmar is presented in Table 4.

General Principles:

1. Inspect so observations are made from the top of the plants.
2. Have the sun at your back or side to increase visibility.
3. Inspect the field to maximize area seen
4. For foundation seed, at least two inspections are required between flowering and harvest.

Table 4:. Field inspection standards for Myanmar

Factor	Foundation	Registered	Certified
Off types (%)	0.05	0.30	0.30
Inseparable other plants(%)	0.01	0.05	0.05
Objectionable Weed (%)	0.01	0.02	0.02
Diseased Plant/Head (%)	0.01	0.05	0.05

Field Selection and Land Preparation

In most countries, fields used for growing certified seed must meet a number of criteria. These include:

1. Not have grown rice for the previous year unless the rice was the same variety planted for certification and met inspection requirements for varietal purity
2. Separation from other fields of same variety by a ditch, levee or roadway or barren strip of at least 3 meters distance must be maintained to preclude contamination through chance out crossing.
3. Thorough land preparation should be done to eliminate volunteer plants and control weeds especially during the fallow or non-cropped period. Fields should be cultivated or chemically controlled with herbicides to ensure that weeds and volunteer rice plants do not seed.

Crop Management

The management of the crop will influence the time and uniformity of crop maturity. Basic requirements of good crop management include good water, nutrient, and pest and harvest management.

Planting Method and Plant Population

Establishing the correct number of plants is essential to maximize water and nutrient use. A target population that results in 4-500 panicles per m^{-2} is desirable. This means establishing at least 70-100 seedling m^{-2} when transplanting, or broadcasting 80-120 kg seed ha^{-1} when direct seeding.

Low populations may result in:

1. increased tillering, which creates more
2. variation in panicle maturity,
3. increased weed populations and
4. reduced yield potential of the variety.

High plant populations may reduce yield and quality by:

1. competing for water and nutrients,
2. mutual shading,
3. lodging and reduced grain size.

Seed Dormancy

Many rice varieties have a dormancy period immediately after harvest. This usually lasts up to 1 month during which time germination levels are low and variable. A number of different treatments are used both to break dormancy and improve seed establishment.

1. Higher temperature. Seeds are exposed to higher temperatures, 40C – 42C, for 1-2 days prior to sowing
2. Seed priming increases the rate of seedling establishment. The seeds are soaked for 4-8 hours and then re-dried prior to sowing. Primed seeds can be sown through a mechanical seed drill but must be sown within 1-2 days after priming.
3. Pre germination where the seeds are soaked for between 24-48 hours then drained for 24 hours before sowing. Pre-germinated seeds have the radicle (root) protruding and can not be sown through a seeder

Water Management

To be able to manage water, the fields must be level and the bunds or levees well maintained. Uniform water depth across the field will contribute to more uniform crop, higher seed yields and consistent moisture content in the grain sample. Reducing the variation in moisture content at harvested reduces the chance of seed damage from disease.

Good water management helps reduce weed competition, which not only increases yields but also improves seed quality by reducing dockage levels and reducing moisture differentials between weed seeds and grain. Wet spots in the seed due to uneven drying or weed seeds can lead to disease and loss of quality.

Nutrient Management

The application of the correct level and type of fertilizer for the variety and growing conditions is essential. The prudent application of nitrogen is essential to get an even maturing crop with full grain size. Excessive and uneven application of N can stimulate late tiller production, which results in heads on the main culm ripening a number of days faster than the tillers. This results in more immature and green heads in the sample as well as higher moisture content that increasing the chance of spoilage by disease. Conversely insufficient nitrogen can lead to reduced grain size and poor vigor.

Roguing

Every field should be rogued to remove off-types, any plants of another crop or variety and diseased plants. Characteristics that can be used to determine varietal purity under field conditions include plant height, pigmentation of plant parts, pubescence, awn characteristics and time of flowering. Roguing should be done at least once before flowering and once after flowering.

SEED PRODUCTION: POST HARVEST MANAGEMENT

Harvesting

The higher the quality of the seed the greater the care is required in harvesting and threshing the seed. Breeder seed should be hand harvested and threshed with a mechanical self-cleaning thresher. Foundation registered and certified seeds can be harvested using machines that have been cleaned thoroughly to avoid mechanical seed mixture.

Timeliness

The optimal stage to harvest grain is between 20-25% grain moisture or when 80-85% of the grains are straw colored and the grains in the lower part of the panicle are in the hard doe stage. This is about 30 days after flowering. If the crop is harvest is too late, many grains are lost through shattering or drying out and are cracked during threshing. Cracked seed will not germinate. If harvested to early, there will be many immature seed grains and this will reduce quality. The immature rice kernels are very slender and chalky and this results in excessive amount of bran and broken grains. Threshing should occur immediately after cutting as the longer the cut panicles remain in a stack the higher the chance of discoloration or yellowing.

Machine Settings

The correct drum speed settings are needed if seed quality is to be maintained and threshing losses minimized. Drum tip speeds for a peg tooth thresher should be between 12-16 m/sec or approximately 600rpm. Higher speed result in higher levels of seed damage while lower speeds increase the amount of seed retained in the panicle. The fan speed and oscillation speed should be between 800-850 rpm. The clearance between the peg teeth and concave should be about 25 mm.

Drying

Rice seed should be dried to less than 14% moisture as soon as possible after threshing. When seed is to be stored for long periods it should be dried to 12% or less and preferably placed in a sealed container.

Table 5: Seed storage life and moisture content

Storage period	Required MC for safe storage	Potential problems
2 to 3 weeks	14 – 18 %	Molds, discoloration, respiration loss
8 to 12 months	13 % or less	Insect damage
More than 1 year	9 % or less	Loss of viability

Drying and tempering the grain a number of times or in stages during the drying process will maintain seed quality. This means drying the grain on a drying pad for a number of hours or in the grain dryer and then tempering the seed by allowing it to cool down for a number of hours in a bin or in the bag. This process should be repeated at least a couple of times until the grain reaches 14% moisture or less.

When sun drying the seed should be spread in thin layers (50-100mm) on the pad or floor and be turned and stirred 7-8 times per day. This will even up moisture distribution and increase the rate of drying. Temperatures on drying pads can exceed 50 C0 during the day and this can cause fissuring. To reduce fissuring the paddy may need to be covered during the very hot times of the day or dried over a couple of days.

Rice is also dried in mechanical batch dryers. The most common smaller dryers have a capacity of 1-3 tons per day with drying times of 6-12 hours. For drying of seed in tropical areas an air temperature of 40-42 C is normally used with a heater capable of raising the air temperature 10-15C. An air velocity 0.15-0.25m/s is required and typical power requirements are 1.5-2.5 kW /ton of paddy. The efficiency of these dryers is also improved by stirring the grain.

Equilibrium Moisture Content

In storage, the final moisture content of seed depends on the temperature and relative humidity of the air that surrounds the grain. The final grain moisture content resulting from storage is called the 'equilibrium moisture content' or EMC.

Cleaning

Threshed seed contains all kinds of trash. This trash can be vegetable, such as chaff, straw, empty grains and foreign seed as well as mineral materials such as earth and stones. Seed should be cleaned as soon as possible after harvesting and certainly before storage.

The simple traditional cleaning method is winnowing, which uses wind or a fan to remove the light elements from the grain.

Mechanical winnowers that incorporate a fan and several superimposed reciprocating sieves or screens are now used in many countries. These can be manually powered or motorized and have capacities from 100 kg to 2-3 tons per hour. Where combine harvesters are use, there is a trend towards using large capacity centralized seed cleaners. These are normally equipped with a series of vibrating sieves and are capable of 10-30 tons per hour.

Storage

If grain is to be stored safely for extended periods it must

1. have less than 13- 14% moisture,
2. be protected from insects and rodents and
3. be restricted from reabsorbing moisture either through rain or the surrounding atmosphere.

When seed is traditionally stored in 20-50 kg sacks it must be periodically fumigated to control insects and in some high humidity environments the seed must be re-dried during the storage period. Some farmers use granaries, which are made from timber or mud/cement or large woven baskets and these are very prone to insect and rodent damage.

Sealed or hermetic storage is an option that has a lot of potential in the tropics. If seed is dried to 14% or less and stored in a sealed storage it reduces the risk of insect and rodent damage and the grain will not absorb moisture from the atmosphere or be damaged by rain. Sealed storage systems can come in all shapes and sizes. They may range from a sealed 200-liter oil drum to the more complex and costly sealed plastic commercial storage. Most large commercial steel and concrete silos being used in western countries can be sealed for fumigation.

Seed Treatments

Seed treatment can be a physical or chemical process.

Physical Seed Treatment

Heat treatments are somewhat rare in rice, but can be accomplished using hot water, dry heat or steam to kill seedborne pests and pathogens. Heat treatment can injury or kill the seed, especially if it is old, injured or has a low heat tolerance. Hot water treatment at 52-57°C for 15 minutes can eradicate seedborne *Aphelenchoides besseyi*.

Chemical Treatment

Fungicides, bactericides, insecticides, and nematicides can be applied as gases, liquids or powder. No broad spectrum fungicides exist, so mixes may often be used. While chemical seed treatment is not widely practiced, the most common chemicals used are Benlate T and Homai. Seed treatement may be used to control pests that are not actually seedborne.

EVALUATION

The use of good quality seed increases crop yields, decreases the number of seeds that need to be sown and reduces the carry over of weeds, insects and diseases. The extent of the yield increase tends to be greater the poorer the quality of the seed that is presently being used by farmers. The question then arises how does a farmer know if using certified or good seed will financially reward him. This can only be answered by doing two key things:

1. Collecting samples of the farmers seed and evaluating the present quality and
2. Testing the field performance of "certified" or "good" versus farmer's seed

Affordability

Based on the cost difference between "Certified", "Good" and "Farmers " seed, the amount of extra grain needed to cover the cost of using improved seed can be readily estimated. For example, in the Philippines, the cost difference between certified seed and paddy is:

Certified seed	= 17.5 P/kg
Farmers/Paddy	= 9.3 P/kg
Difference	= 8.3P/kg

The extra harvest required to cover the cost of certified seed, based on different seeding rates is given in the following table.

Table 6: Yield increases needed to cover extra seed cost.

Seed rate (kg/ha)	Extra cost (P/ha)	Extra yield needed to pay for seed (kg/ha)
40	328	35
80	656	71
120	984	106
160	1,312	142
200	1,640	177

If the 'yield gain' were 5%, then on a 3 t/ha yield, the yield gain would be:

$$3{,}000 \times 0.05 = 150 \text{ kg/ha}.$$

This means that a farmer who presently plants at 160 kg /ha would be better off using certified or good seed rather than paddy or farmer seed. With better quality seed, sowing rates and pest problems will also be reduced and therefore the total returns will be even greater.

METHODS FOR EVALUATING GOOD SEED

Moisture Content

There are two methods that can be used to measure moisture content:

The primary or direct method, often called the drying oven method. This method weighs a sample of the paddy and then removes all the moisture by heating the sample in an oven. After the moisture has been removed, the sample is re-weighed.

The secondary method uses electronic moisture testers. This method uses meters that measure the moisture content either by measuring the resistance or capacitance of a sample of the paddy.

Drying Oven Method

The drying oven method takes longer to perform but provides the most reliable measure of moisture content.

1. Set the oven at 130°C.
2. Weigh three 100-gram seed samples [A].
3. Place the three samples inside the oven and leave for 16-24 hours.
4. Measure the final weight of each sample after the 16 to 24 hours [B].
5. Compute the moisture content for each sample using the equation:

$$\text{MC} = \frac{Initial\ Weight[A] - Final\ Weight[B]}{Initial\ Weight[A]} \times 100$$

6. Compute the average over three samples to obtain a reliable measure of the moisture content.

Electronic Moisture Meters

1. Read the operator's instructions
2. Turn the moisture meter on
3. Ensure the machine is set for paddy
4. Fill the tray or bowl of the moisture tester with a sample of the paddy to be tested
5. Turn or press the knob until the moisture reading is displayed
6. Test at least three samples and calculate the average of the three readings

Moisture meters are most reliable after they have been calibrated. To do this, use the drying oven method to determine an accurate moisture content reading for a sample of paddy. Test a sample from the same batch of paddy using the moisture meter and compare the results.

Germination Test

A germination test is often the only test a farmer can conduct on the seed before planting. Monitoring the time taken to germinate will also give an indication of vigor. This procedure is easy, inexpensive and portable.

Sampling

To obtain a random sample for testing it is always best to take samples from different parts of the bag or container. If the grain to be tested is from a seed lot contained in more than one bag, sample must be taken from several bags. A good rule of thumb for determining how many bags to sample is to take samples from a number of bags that represents the square root of the lot size. For example, if the lot contains nine bags, then sample at least three bags. If the lot contains 100 bags, then sample from at least 10 bags.

Equipment

To conduct this test you will need the following:

1. Waterproof tray. A flat-sided water bottle, cut in half lengthwise makes a good tray.
2. Water absorbent material. Tissues or cotton wool are ideal.
3. Seeds.
4. Water supply.

Procedure

1. Place water absorbent material inside the waterproof tray
2. Take random sample from each seed lot and mix in a container
3. Take at least three seed samples from the mixed grain
4. Count out 100 seeds from each sample and place on absorbent material inside the tray
5. Carefully saturate the absorbent material
6. For each of ten [10] days

   Each day check that absorbent material remains moist

   Record the number of germinated seeds
7. Compute germination test for five [5] days and for ten [10] days The rate of germination is an indicator of vigor. Rapid seed germination increases the chance that seed will establish in the field.

Calculating the germination rate

Germination rate is the average number of seeds that germinate over the five and ten day periods.

$$\text{Germination (\%)} = \frac{\textit{Number of seeds germinated}}{\textit{Number of seeds on tray}} \times 100$$

For example, If 86 seeds germinated in a tray of 100 seeds after 10 days, then

$$\text{10-day germination (\%)} = \frac{86}{100} \times 100 = 86\%$$

Seed Lot Purity

The impurities measured to test the purity of the seed batch are

1. weed and other crop seed, and
2. inert matter.

Procedure

1. Select at random a sample of the seed. 100gm is a good sample size.
2. Weigh the sample **[A]**
3. Remove all of the weed and other crop seeds and the inert matter
4. Weigh the weed and other crop seed matter removed. **[B]**
5. Compute the weed percentage as follows:

$$\text{\% Weed} = \frac{\textit{Wt of weed}}{\textit{Total wt of sample}} \times 100$$

6. Weight the inert matter removed from the sample **[C]**, the inert matter percentage is calculated using the formula:

$$\% \text{ Weed} = \frac{Wt\ of\ Inertmatter\,[C]}{Total\ wt\ of\ sample\,[A]} \times 100$$

Varietal Purity

Grain Dimensions

The grain size and shape (length-width ratio) is a very stable varietal property that can be used to measure the varietal purity of a sample. Comparing the length-width ratio of the sample with the published ratio for the variety indicates the varietal purity of the grain sample. Significant deviation indicates that the sample is impure – that is, it is either a different variety or a mixture of varieties.

Procedure

1. Obtain a random sample from the seed batch.
2. Collect twenty grains at random from this sample of seed
3. Use a Vernier caliper or photographic enlarger to measure the dimensions of each of the twenty grains.
4. To obtain the seed shape, the length width ratio, the following equation is used

$$\text{Length to Width ratio (L/W)} = \frac{Average\ seed\ length\,[A]}{Average\ seed\ width\,[B]}$$

1000-grain Weight

A second characteristic that measures varietal purity is the 1000-grain weight. Each variety has a published weight for 1000 grains. If the 1000-grain weight calculated from the sample departs from this, it may be an indication that the sample contains a mixture of varieties.

Procedure

1. Select a random sample from the seed batch
2. Count 1,000 whole grains from the sample.
3. Weigh the 1,000 grains.

   This is the 1,000-grain weight.

Red Grains

A grain is considered to be red if more than 25% of its surface area is red colored or streaked.

Procedure

1. Select a random sample of the milled rice. 25gm is a good sample size.
2. Weigh the sample **[A]**.
3. Select and separate the red grains from the sample. The red kernels are those that have 25% or more of the grain red as shown in the diagram below:
4. Weigh the red grains separated from the sample **[B]**.
5. Calculate the percentage of red grains in the sample using the formula:

$$\%\text{ red grains} = \frac{Wt\ of\ red\ grains\,[B]}{Total\ weight\ of\ sample\,[A]} \times 100$$

SEED HEALTH TESTS

Fungal Contaminants

The primary seedborne fungi causing yield loss are: Blast (Pyricularia oryzae CAV.), Brown spot (Helminthosporium oryzae Breda de Hann) and Bakanae (Fusarium moniliforme Shel).

Blotter procedure:

Blotter and agar plate methods are a simple and inexpensive way to detect seedborne fungi that respond to sporulation.

1. Use 9.5 cm Petri plates made of pyrex glass or clear plastic to allow NUV light to penetrate. In each dish, place 2 or 3 layers of good quality white or colored blotting paper moistened with distilled water.
2. Place seeds from the sample (with or without pre-treatment) equidistantly on the petri plate at 25 seeds/plate.
3. Incubate seeds at 22°C under 12-h light and 12-h dark cycles with NUV light.
4. Express results as a percentage of the number of total seeds.

Agar plate procedure:

Detects seedborne fungi through colony characteristics exhibited as they grow on the agar

1. Take 400 seeds as a sample. Pretreat seeds with 1% sodium hypochlorite for 10 minutes.
2. Drain off excess liquid. Place seeds (10 per agar dish) on the agar plates (in 9.5 cm Petri dishes)
3. Incubate at 22°C for 5-8 d, either under alternate cycles of NUV light and darkness, or in darkness.

Identification

1. Prepare slide mounts of spores or other bodies in a drop of water or lactophenol blue and examine under a compound microscope for shape, size and color.
2. Compare results with published data.

Bacterial Contaminants

Two species of bacteria – Pseudomonas and Xanthomonas are the primary bacteria of concern. Seed borne Xanthomonas can cause Bacterial leaf blight. Bacteria have to be extracted, purified, isolated and then identified.

Nematodes

Only one of the more than 100 nematode species (Aphelenchoides besseyi) is known to to seed borne. Nematodes are unsegmented round worms. They are micropcopic and transparent. They can live in the roots or in aerial plant tissue. For practical seed testing, the "modified Baermann funnel method" is considered suitable. Seeds are soaked and incubated before nematodes are counted in the water extract.

Viruses and Mycoplasmalike Organisms

Of the 21 rice viruses and mycoplamalike organisms, only two (wrinkled stunt vitrus and withches' broom virus) are confirmed as seed transmitted. They have only been confirmed in Surinam and seed health testing has not been established.

VEGETABLE SEED QUALITY

Some vegetables are harvested for market-use such as carrot, onion, radish, brassicas, and lettuce are typically grown in areas with lower temperatures from those required for a seed crop in order to maintain good vegetative growth and minimise bolting. Seed production of these crops often requires higher temperatures to encourage bolting although extremely high temperatures also hinder pollen growth and fertilisation which lead to reduced seed number per floret or fruit. This is true in onion, tomato, and lettuce. In contrast, beet and carrot market-use crops have been selected to avoid bolting in mild climates but they still require adequate periods of cool temperatures (vernalisation) for flowering and seed production.

The Pacific Northwest of the United States provides such sustained cool temperatures and this is the reason why beet and carrot seed production occurs in this region while market-use production of these crops is geographically more diverse. The requirement of some vegetables for specific photoperiods before induction of flowers also limits those areas where vegetable seed production is successful. This effect is dependent on the latitude and the length of the photoperiod which greatly affect seed yield. For example, onion is a vegetable crop that exhibits a wide diversity in photoperiod response and this influences where seed production occurs.

Seed production of long-day onions occurs at latitudes greater than 45o N and short-day onions at latitudes no greater than 35°N. Limited seed production also can be accomplished in glasshouses where supplemental light is provided or light excluded to promote flowering. Even when appropriate photoperiods are present, temperatures must be suitable for the completion of the entire life cycle of the crop during the growing season. Successful market-use crops such as onion, radish, broccoli, and cabbage are harvested before flowering occurs. As a result, these crops are grown in regions with cooler temperatures not suitable for seed crops which require longer periods and warmer temperatures for maturation of seed.

Precipitation amount and distribution also varies for market-use and seed crops. Many vegetable crops harvested for their vegetative structure require large amounts of early season rainfall for most rapid development. Seed crops, in contrast, demand uniform precipitation during development followed by dry conditions during seed maturation. Under these environments, dependable conditions for seed harvest are possible and pathogen infestations minimised. This is the principal reason much of the vegetable seed production has shifted to the western United States. The Mediterranean climate of the Pacific Northwest assures cool, dry summers and warm, wet winters for optimum development of many biennial seed crops. Southern California and much of Arizona are hot and dry during the summer which is ideal for vegetable seed production. The dry condition of these locations permits seed crops to be irrigated with the proper amount of moisture at the proper time so that optimum seed quality is achieved.

The following represents general requirements and considerations utilised in vegetable seed production.

Tillage

The preparation of a soil site for vegetable seed production differs little from that for fresh market production. Soils should have a high water holding capacity to encourage uniform vegetative growth. Depending on the crop and water availability, planting is done on raised, shaped, or flat planting surfaces which require appropriate soil preparation. In all cases, the seed bed must be uniform because most vegetable seed crops are precision planted and uniform emergence and seedling development necessary for optimum performance.

Planting

The planting of vegetable seeds for seed production is generally done by direct drilling or transplanting of glass house grown seedlings. Biennial vegetable seed production, however, can be achieved using two planting techniques. The "seed-to-seed" method is done by planting biennial seed, allowing the resulting plant to overwinter without being transplanted, and harvesting the seed crop the following season. This method does not permit the selection or roguing of root or other genotypic characters. A majority of carrot and onion seed crops are produced by this method which is less expensive and complex than the "root-to-seed" method.

The "root-to-seed" method is done by planting biennial seed, removing the resulting plant from the soil (in some cases stored), replanting, and harvesting the seed crop the following season. The replanted plant is referred to as a "steckling." This method permits the selection or roguing of root or other genotypic characters. It also decreases the time required for varietal development. For example, onion bulbs adapted to dry regions can be planted in the spring in the desert regions of the southwest and then shipped for seed production to central California or the Pacific Northwest for planting in the early fall of the same year. This is the most expensive method of vegetable seed production due to transportation, handling, and replanting costs, but the benefits of checking for minimum genetic quality standards before replanting are important in vegetable seed production.

Vegetable seed crops must be planted so that sufficient vegetative development occurs to support optimum fruit and seed development. In addition, the increasing emphasis on hybrid seed production necessitates planting of male and female lines at times that assure successful pollination, a process known as nicking. This may require that genetically different male and female lines be planted at differing times so that synchronous flowering occurs. Most hybrid broccoli is transplanted and staggered plantings are necessary for the male and female lines to ensure proper nicking. Hybrid cauliflower and squash are direct seeded but because of the differences in time required for the male and female lines to flower, they are planted in a relay fashion where the earlier flowering line is planted first followed by the later flowering line.

Vegetable seed production also is unique from most agronomic crops because vegetables such as beets, carrots, and cabbage are biennials and must develop sufficient vegetative growth prior to cool temperature exposure in order that vernalisation successfully induces flower formation the following season. Planting such crops too early causes winter kill or late season pest infestations. Planting too late results in a lack of vernalisation which limits flowering and reduces seed yield. In some instances, vegetable seed producers opt not to start the seed crop by seed. Rather, to ensure optimum vegetative growth for vernalisation, they plant some developed portion of the vegetative structure. This is commonly done with carrot and beet stecklings and onion bulbs which allows for maximum production of plant biomass before vernalisation.

Vegetable seeds also vary greatly in their tolerance to soil temperatures at planting. Pea, radish, and spinach require cool soil temperatures for optimum seedling emergence. Beets, cabbage, carrot, and onion are tolerant of cool soil temperatures although they do better in warmer soils. Squash and melons require warm soil temperatures for optimum seedling emergence. In regions with high summer temperatures, vegetables such as tomato, pepper, eggplant, and cucurbits are planted in the early spring to optimise flowering and seed set in early summer. Later plantings of these seed crops are possible by transplanting young plants previously started in glass houses. This avoids the delayed establishment of plants from direct seedings due to cooler night temperatures and wetter soils encountered in the early spring.

Row spacings and planting densities of vegetable seed crops differ from those for fresh market production. Sufficient space for flower development, air movement to reduce pathogens, unrestricted access to inflorescences by pollinators, mechanical cultivation, and harvest operations are necessary for seed crops. In some cases (e.g., eggplant, pepper, tomato, muskmelon, watermelon), row spacings used for seed production are the same as those for fresh market production. In most cases, row spacings are different for vegetable seed production.

In lettuce, the size of the head will be larger than the fresh market head in order that bolting is promoted. Large-head types are planted 25 to 30 cm apart while small-head and loose-leaf types are planted 15 to 20 cm apart. Similar considerations are encountered for cabbage. Broccoli and cauliflower produce sizable seed stalks requiring rows 40 to 60 cm apart. Planting densities are often different as well in vegetable seed crops from fresh market vegetable crops. Beet seed production is accomplished at higher planting densities than beets harvested for their roots. Carrot, celery, and parsnip seed production is planted at densities much less than for root or stalk production. Higher planting densities of onions grown by the bulb-to-seed method reduces lodging of the large seed stalk scapes.

The method (wind or insect) of pollination affects row spacings and planting densities. Vegetables that are wind pollinated include sweet corn and spinach. Insect pollinated vegetables include most of the cole crops, carrot, and onion. Row spacings and planting densities are also important considerations in the production of vegetable seed hybrids. In order to maximise the amount of hybrid seed produced, the optimum ratio of female to male rows must be determined.

Fertilisation

Vegetable seed crops have similar nutrient requirements as fresh market crops. Where possible, nutrients that stimulate reproductive development can be supplemented with normal fertility regimes. Boron, for example, increases the number of fruiting sites and seed produced per pod in legumes such as pea and bean. Molybdenum also is essential for optimum nitrogen fixation of legume seed crops. Another approach to stimulating reproductive development in vegetable seed crops is to split the application of fertilisers at planting and before flowering. This avoids luxury consumption of the elements at the time of application, improves crop uptake efficiency, and increases the formation of reproductive structures. Generally, lower nitrogen and phosphorus levels are used in seed production because these encourage continued vegetative growth. Because most vegetable seed crops are grown in rows, fertilisers are typically applied either in bands in the planting row or as a side-dressing beside the row after establishment.

Weed and Pest Control

The management of weeds and pests in vegetable seed production poses additional problems due to their longer growing season not encountered for vegetables grown for market use. This extended growing period necessitates an increase in residual pesticide activity or additional pesticide applications. In some cases, governmental regulations do not permit the use of the same pesticide for a vegetable seed crop that is used for the market use crop. This is because the approval process is directed at the more economically important value crop.

Some pesticides which control insects also detrimentally affect insect pollinators essential for cross pollination or hybrid seed production and cannot be used in a seed production field. Such restrictions on the use of pesticides in vegetable seed production imply that the safe use of pesticides will be more difficult in the future and greater reliance on crop rotations and biological control will be necessary.

Insects and the timing of their infestations affect vegetable seed production. The Lygus bug prefers legume crops and safflower as its primary host. However, once these crops are harvested, it commonly flies to adjacent carrot seed fields and rapidly creates epidemic infestations. There, it feeds on developing seed embryos and subsequently reduces seed yield and quality. Cabbage looper is a significant pest of brassicas during vegetative development but does not affect seed stalk and pod development. Thus, control of this pest is important during vegetative growth if maximum yield and seed quality are attained. Aphids affect the harvest and subsequent conditioning of brassica seed crops by the development of "honeydew" which adheres to the seed pods and makes separation difficult.

Vegetable seed crops are susceptible to the same diseases as fresh market crops. However, the need to complete their entire life cycle for seed production means that other diseases may also be encountered and require control. Many significant diseases of vegetable seed crops can be controlled by growing the crop in differing geographic locations. For example, the low relative humidity and high temperatures of the southwestern United States reduce the incidence of the seedborne pathogen *Cercospora carotae* in carrot seed fields commonly found in the high relative humidity, low temperature Pacific Northwest. The incidence of downy mildew is reduced by growing long-day onion seeds in the low humidity areas of the desert southwestern United States.

Diseases in vegetable seed fields can also be minimised by separating the seed crop from disease pests either physically or over time. Physical separation occurs by producing the seed crop in an isolated region from fresh market crops or from the disease causing organism. Celery seed fields are isolated from celery stalk production fields to reduce the incidence of late blight caused by *Septoria*. Lettuce seed fields are isolated from leaf and head production fields to reduce the incidence of seed-borne lettuce mosaic virus.

Isolation from the disease causing organism is achieved by maintaining weed-free irrigation canal banks and roadsides which serve as alternate hosts or growing the seed crop where the disease causing organism does not exist. To meet phytosanitary requirements for export, onion seed crops must be grown in regions where certain disease and nematodes are certified to be absent. The lettuce mosaic virus is another serious disease that results in unmarketable lettuce heads. To control this disease, an integrated pest management strategy has been established which focuses on the planting of disease-free seed and control of insects and weeds. Disease-free seed is achieved by planting stock seed in glass houses free of such virus vectors as the green peach aphid (*Myzus persicae*). Commercial seed lots are certified to have not more than one infected seed in 1,000 seeds that have been screened. Other effective measures include control of common weeds that serve as alternate hosts to insect vectors and frequent roguing of lettuce seed fields to remove any plants suspected of carrying the virus.

Some disease causing organisms have relatively short life spans. In these cases, separation of the seed crop from the organism over time reduces disease problems. Alternating cabbage and other cruciferous seed crops in a three year rotation with non-cruciferous crops decreases the incidence of club root caused by *Plasmodiophora brassicae*. Nematodes can similarly be controlled in cucurbit and solanaceous seed crops by three year rotations with crops not from these families. In other cases, diseases can be controlled by storing the seed for a period during which the disease causing organism dies. An example is soft rot (*Erwinia carotovora*) on celery seeds which dies after two years of seed storage. Hot water baths or chemical treatments also control diseases. Hot water treatment is used on a variety of vegetable seeds to control a number of seedborne bacteria and fungi. Many disease causing organisms are carried on the seed coat and most of these are controlled by various topical seed treatments consisting of fungicides, insecticides, and bactericides.

Harvesting, Threshing, and Drying

Vegetable seed crops are divided into two categories based on the type of fruit harvested. Dry-seeded fruits include the brassicas, legumes, and onion and wet-seeded fruits include the cucurbits, melons, and tomatoes. The methods of harvest and extraction of seeds vary dependent on the type of fruit harvested. Of great concern in dry-seeded fruit vegetable seed production is the potential for loss of yield due to seed shattering when harvesting occurs after physiological maturity. To minimise seed loss, two approaches are employed.

Since many dry-seeded vegetables achieve physiological maturity before harvest maturity, they are cut while the plant is still green and placed into rows where the wind further dries the plant and its reproductive structure. This permits more uniform maturation of all dry-seeded fruits and seed quality is enhanced compared to the more costly option of sequential harvest of mature fruits. Later, after the seeds have reached harvest maturity, they are extracted from the seed heads either by hand or mechanically using a combine. The second approach is to visually monitor the development of the dry-seeded fruit and harvest the crop before shattering occurs. For example, onion seeds are very susceptible to shattering once the florets have dried and seed germination and quality is reduced if the scape stalks are cut too early.

A common harvest criterion has been to examine umbels for the first mature seeds. In indeterminate crops such as carrot, seeds mature in sequential order with those on the primary umbel maturing first followed by those on secondary and tertiary umbels. Secondary umbels produce about 50% the quantity of pure live seed of the primary umbel. Thus, harvest of this dry-seeded crop is complicated by the umbel maturation sequence. Seeds produced from the first lettuce flowers also tend to be larger and of higher quality than seeds produced from later flowers.

Seed harvest from wet-seeded fruits is often determined by fruit color. Seed quality in pepper as determined by percentage germination and rate of germination is higher in seeds harvested from red fruits compared to those harvested from green fruits. In tomatoes, maximum seed germination is attained when the fruits are at a mature green stage compared to those which are at the red or fully-ripe stages.

After the fruits are determined to be at the correct stage for maximum seed quality, they are gathered and the seed extracted. In some cases, this is done in one mechanical operation where the fruit is picked from the row, conveyed to a crushing mechanism, and the pulp removed from the seed by washing with water in a rotating drum. For high moisture vegetables such as melons and tomatoes, water is periodically drained from the equipment. For low moisture vegetables such as squash and peppers, water is periodically added to the equipment.

In most cases, vegetable seed companies mechanically harvest crops such as tomatoes, peppers, and cucumber in the field, transport the fruit to a central extraction plant, and remove the seeds from the fruit. In tomato, cucumber, and some melons, a mucilaginous layer around the seed must be removed prior to cleaning. This can be done by two methods. The first, and most common, is the retention of seeds in the crushed fruit pulp for a few hours to more than a day so that fermentation occurs. This process is closely monitored because the fermentation creates heat and mechanical injury to the seeds if continued too long. The second approach is to add sodium carbonate or hydrochloric acid to the crushed fruit extract to degrade the mucilaginous layer. After the fruit is crushed and any mucilaginous layer removed, the seed is separated from the pulp and passed along a water sluice containing gates along its entire length.

Mature high quality seeds sink behind the gates and are separated from the pulp by the running water. After the pulp is removed, the sluice gates are opened and the mature seeds captured on a screen. At this stage, the wet seed must be promptly dried to avoid heating and loss of seed quality with increasing imbibition. To accomplish this, the wet seeds are collected and placed into a spin dryer to remove excess water and the moist seeds are evenly spread on screens and exposed to large volumes of heated air. After drying is completed, the seeds are handled and conditioned in the same way as employed for dry-seeded vegetable crops.

Cleaning and Storage

The physical quality of the seed lot following harvesting, threshing, and drying must be evaluated to determine whether precleaning is necessary. If the seed lot possesses large amounts of inert material such as stones or plant debris, it should be passed over a scalping machine. If the seed possesses undesired appendages which cause the seeds to clump together such as carrot, then it must be passed through a debearder to ensure free-flow of individual seed units. Impact debearders are also used in multigerm beet seeds to split the fruit into single seed units. After precleaning, vegetable seed crops are conditioned in the same way as other seed crops using the same principles and equipment.

Some additional improvements can be made. For example, greater seed density improves the seed quality of lettuce, tomato, and onion crops. This can be accomplished by placing the seed mass into a solution of specific osmotic concentration. The more dense seed fractions are retained and the more buoyant seeds removed. After cleaning is completed, vegetable seeds are stored using the principles of successful seed storage.

Rather, six of the most important vegetables (onion, tomato, cabbage, lettuce, cucumber, carrot, and sugar beet) are considered. These were selected not only because of their economic value but also because they represent differing botanical families, generally differ in their vegetative/ reproductive structure harvest, and illustrate differing seed production practices. Similarities exist in the seed production of related species and many of the principles detailed here can be extended to these crops as well.

Onion (Allium cepa)

Onions originated in Iran and Pakistan and were established staple foods of Egypt and India around 1500 B.C. By the middle ages, they were in Europe and brought to North America by Spanish settlers.

The unique flavor and odor of onions have made them an excellent food source that provides zest to many dishes. Moreover, the numerous ways in which onions are prepared from boiling, frying, stewing, baking, pickling, to eating raw make them a versatile food source. The recent popularity of a health conscious world for salad bars has further increased their agricultural importance. Nutritionally, onions are low in calories (approximately 40 calories for an average size onion) and high in ascorbic acid. They are classified into three agriculturally important groups: common, aggregation, and proliferous. The common onion is the most agriculturally important group. It is a bulb producing type reproduced by seed. The aggregation group is characterised by formation of many lateral bulbs or shoots and is reproduced vegetatively. An example of this is shallots. The proliferous group has bulblets form on the inflorescence and is vegetatively reproduced. An example of this is top-set onions.

PLANT DEVELOPMENT

Vegetative. Onions are a member of the Liliaceae family and are considered a cool-season crop. They are herbaceous biennial plants. The first year of development results in the establishment of a fibrous, shallow root system attached to a base plate of fleshy leaves arising around a small conical terminal stem, a structure botanically termed a bulb. Because the root system is shallow, onions are particularly prone to dehydration injury. The leaf sheaths that project above the ground are sometimes called a false stem to distinguish them from the true stem at the bulb's base. Bulb formation is initiated by changes in day length. Very early types require a 12 hour photoperiod and later types initiate bulbing with a 15 hour photoperiod.

Reproductive. Flowering and seed formation occur in the late Spring or early Summer in response to a vernalisation treatment during the Winter. The onion inflorescence is an umbel that produces 50 to 2,000 flowers. Flowering can be as long as two weeks and is not uniform since the umbel actually consists of aggregations of smaller 5 to 10 flowered inflorescences called cymes. From one to over a dozen flower stalks can be produced on a single plant dependent on its development. More flower stalks are produced on plants grown from bulbs than those grown from seed. The stigma of the flower becomes receptive to pollen after the anthers have released their pollen thereby reducing self-fertilisation. However, open-pollination primarily by insects such as honey bees and leafcutter bees carrying pollen from one flower to the next is common.

SEED

The seed is three sided, 3.0 to 3.5 mm (0.12 to 0.14 inches) long. Its surface is wrinkled, dull and greasy looking, black, and has a depressed scar at one end. Within the seed, the embryo is 6.0 mm (0.24 inches) long and 0.4 mm (0.02 inches) in diameter and is curved in a complete circle. Most of the embryo is comprised of a single cotyledon to which is attached a very small shoot apex, first foliage leaf, and short primary root. It is embedded in a tough thick-walled endosperm. The 1,000 seed weight of onion is 3.6 g (126,000 seeds per pound).

SEED PRODUCTION

Onion seed production can be open-pollinated or hybrid. Since hybrids have greater uniformity, increased yields, and greater disease resistance, most seed produced today is hybrid. These can be produced either by a bulb-to-seed or seed-to-seed method. The bulb-to-seed method is when onion bulbs are grown, harvested in the fall, stored, and replanted in the Spring. The cold vernalisation requirement is experienced during bulb storage in the Winter. This method is more expensive than the seed-to-seed method but it also permits a grower to easily discard off-types, diseased or otherwise undesirable bulbs. It is practiced primarily for maintenance of onion stock seed.

The seed-to-seed method is when seed is planted in the Summer, carried over through the Winter as bulbs in the field, followed by flowering in the Spring. This method results in higher seed yields than the bulb-to-seed method because there are more plants per hectare (acre) and less time is invested in bulb establishment. While overall costs are less than the bulb-to-seed method, careful attention must be given to pest control. Roguing programs to eliminate off-type plants are also required.

Tillage. An onion field should not be used for seed production more than once every four years. The field must be free of perennial weeds and soilborne diseases. The soil should be friable, fertile, well supplied with humus, and well drained. The soil should be plowed to a depth of 15 to 20 cm (6 to 8 inches), harrowed, and worked until a uniform firm seedbed 10 to 15 cm (4 to 6 inches) deep is established.

Planting. Onions seeds are planted 1 to 3 cm (1/2 to 1 inch) deep in rows 50 to 75 cm (20 to 30 inches) apart dy drills at a rate of 4.5 to 6.7 kg per hectare (4 to 6 pounds per acre). Higher seeding rates and closer row spacings have, in some cases, led to higher yields. Planting should occur as soon as the danger of frost is over. If the fields are irrigated or rainfall is common, row spacings of 30 to 45 cm (12 to 18 inches) can be used.

Fertilisation. The objective of fertilising onion seed crops is to ensure that the onion bulbs reach adequate size before the onset of Winter in order that they can produce vigorous seed stalks the following Spring. Onions respond well to fertilisers because the root structure is limited and shallow. They grow best on soils that are slightly acidic (pH 6.0 to 6.5). Mineral soils should be amended with organic matter to assist in moisture retention and improve the soil physical structure. The equivalent of 56 kg per hectare (50 pounds per acre) of nitrogen should be residually available in the soil or the equivalent applied broadcast before planting. Additional applications of nitrogen per hectare (acre) can be applied in one or two sidedressings to make it easily available to the roots. Both phosphorus and potassium should be broadcast at the rate of 67 kg per hectare (60 pounds per acre) prior to planting. Later applications, if needed, can be sidedressed. Fertilisation of nitrogen and potassium is not recommended at flowering because these make the flower nectar unattractive to bees.

Minor element deficiencies occur in onion, particularly on muck soils. Copper deficiencies result in thin, poorly colored scales that have a poor bulb storage life. This can be corrected with the application of 224 kg per hectare (200 pounds per acre) of copper sulfate. This treatment lasts for several years. Magnesium deficiencies result in leaf chlorosis and are often encountered on muck soils that are alkaline or have recently been limed. Manganese deficiencies cause a gradual interveinal yellowing, twisting, curling, and stunting of leaves. Manganese sulfate at 168 kg per hectare (150 pounds per acre) mixed with the broadcast fertiliser alleviates this problem.

Irrigation. Seed onions produced in the dry West require irrigation because of their shallow roots and high demand for water. An onion crop can use 1 1/2 to 2 1/2 acre feet of water during its development and maturation. Soil moisture should always be above 65% field capacity. In areas where irrigation is practiced, seedbeds are usually irrigated immediately after planting. Then, one to three irrigations are necessary after seedling emergence to ensure rapid, continued growth. As the plants initiate bulbing, irrigation is stopped and the soil allowed to dry.

Weed and Pest Control. Onions are poor competitors against weeds because of their slow growth, small stature, shallow roots, and lack of dense foliage. Weeds interfere with harvest and can become contaminants of the crop seed. The most effective weed control measures include crop rotations, cultivation, and use of selective herbicides. A short fallow period before planting also permits an opportunity to use tillage and non-selective herbicides to control weeds. Cultivation is one of the most successful weed control techniques in onion. Weeds can be destroyed during seedbed preparation. After planting, cultivation every one to two weeks until bulbing occurs is useful. This practice not only suppresses weeds but also loosens the soil after it has hardened following irrigations and rains. Initial cultivations should be shallow since the onion roots are close to the soil surface and damage to the crop can occur.

Harvesting, Drying, and Threshing. Although all seed heads on an onion plant do not mature simultaneously, there still is usually one harvest in the seed field. The seeds are dry enough for threshing when the capsules and small seed stems are brittle and readily break when rolled in the palm of the hand. Threshing is done with a combine.

Conditioning. After threshing, onion seeds are conditioned to eliminate weed seeds, light seeds, and chaff. Initial cleaning is with an air-screen cleaner followed by a gravity table. In some cases, the light separated seed is washed. Heavy, good seeds sink while light, poor seeds float. The washing treatment should not exceed three minutes since the seeds rapidly absorb the water and quality is subsequently reduced. The heavy seeds are retained and immediately spin dried in a centrifuge and further air-dried.

Storage. Onions possess one of the most rapidly deteriorating seeds among major crops. They quickly lose complete viability in less than a year when stored under hot, humid conditions. If properly stored, however, viability of onion seed can be retained for long periods and low seed moisture content and low storage temperature favor longer storage life.

Cabbage (Brassica oleracea)

Most wild forms of cabbage are found along the coasts of the Mediterranean Sea which is considered their point of origin. Greek and Roman civilisations were the first to recognise the value of cabbage. Its ease of growth and excellent storability ensured its place as an important crop. Initially, these cultures used cabbage for medicinal purposes. Cabbage juice was employed as a gargle against hoarseness and the leaves covered wounds and ulcers to hasten healing.

There are three major types of cabbage classified according to market demand: fresh market, storage, and sauerkraut. Fresh market types usually form conical to globe heads and have a range of maturities from 50 to 95 days. Storage types are slow in growth and have a firm head with thin, finely veined leaves. Cabbages are used in a variety of ways from raw in salads and soups, boiled or steamed in mixtures of vegetables and cabbage rolls, to pickled in sauerkraut.

PLANT DEVELOPMENT

Because cabbage is a biennial plant, the first year of growth requires the production of a vegetative head followed by flowering and seed production in the second year. It is considered a cool-season crop which grows best between temperatures of 10 and 25C (50 to 77°F).

Vegetative. The cabbage seedling often produces a red hypocotyl with two notched cotyledons and a tap root with lateral roots. The first three leaves have a petiole while later leaves are completely sessile. The leaves are unique from many other crops because they are coated with a layer of wax. The first leaves unfold normally but later leaves only partially unfold and do not expand as much as earlier leaves ultimately forming a head that varies from flat-topped to long-oval in shape.

Reproductive. Cabbage is not photoperiodically sensitive but the apical meristem of plants that have begun to form a loose head must be exposed to four to six weeks of 4 to 7C (39 to 45°F) for vernalisation to occur eliciting production of a reproductive apical meristem. Vernalisation of the seed or seedling does not cause flowering. The flowering process starts from the bottom of the raceme and moves to the top. This often results in racemes on which mature siliques are found on the bottom with unopened flower buds on the top causing an uneven setting and ripening of seed on the plant. The flowers are insect pollinated. When hybrid seed is produced, this normally occurs in a 2:2 ratio of male:female parent rows. After pollination, elongation of a dry pod called a silique containing 10 to 30 seeds is completed after three to four weeks. Dehiscence occurs when the two silique valves split from the point of attachment to the plant upward leaving the exposed seeds attached to the placenta.

SEED

The cabbage seed is almost round, 2.0 to 3.0 mm (0.08 to 0.12 inches) in diameter, reddish brown to grey to black, with a discernable radicle groove and hilum. The seed coat possesses a small reticulated surface. The embryo is large, folded and little endosperm is present in the mature seed. The 1,000 seed weight is 3.6 g .

SEED PRODUCTION

Tillage. Cabbage requires an abundant supply of water. As a result, it typically grows best on soils that are heavy and contain a high level of organic matter. The pH should be between 6.0 to 6.5 for optimum growth. The planting bed for cabbage should be deep and well worked to enable rapid seedling root development and access to water. This can be accomplished by disking the soil. Disk cultivators are useful in preparing seed bed rows about 7 to 10 days before planting because they further pulverise the soil. Additional smoothing of heavy soils, if required, can be done with a harrow.

Planting. Normal cabbage seed production begins with planting in the

Summer followed by the half-grown plants over wintering, flowering in the Spring, and maturation and harvest of the seeds in the Summer. All stock seed is tested or hot water treated for black rot prior to planting. Cabbage is planted initially in beds or the greenhouse. The seedlings are then transplanted 45 days later in the field in rows 107 cm (42 inches apart) with a spacing between plants of 40 cm (16 inches) to produce a population of 3,645 to 4,050 plants per hectare (9,000 to 10,000 plants per acre).

Hybrid seed production is more labor intensive. Each parent is normally staked and stringed to allow improved pesticide spraying and facilitate subsequent hand cutting of the inflorescence. The stakes and strings insure that the parents stay separated. The nick between female and male parents must be managed to minimise the number of inbreds and maximise the number of hybrids produced. To accomplish this, bees are placed in the field at 5% bloom. In some cases, plants are topped if one parent comes into bloom earlier than the other.

Fertilisation. Nitrogen during the establishment of cabbage plants leads to a softening of tissues and an increased susceptibility to frost. As a result, nitrogen is usually omitted at planting with phosphorus and potassium applied at rates of 100 and 150 kg per hectare (90 and 134 pounds per acre), respectively. If nitrogen is necessary, it is split as a sidedress in the Fall and added in the Spring just before bolting. This ensures maximum growth of the sidebranches which enhances seed yield. Too much nitrogen in the Spring or Fall enhances soft tissue development, increases the risk of lodging, and promotes extended flowering which may cause increased inbreds in hybrid seed production. Cabbage is not as sensitive to micronutrient deficiencies as are other members of the Brassicaceae.

Weed and Pest Control. Chemical and mechanical weed control programs are practiced with cabbage. Seedbed preparation should eliminate most weeds. When weeds are persistent, however, a preplant or preplant incorporated application of herbicides can be used. The extensive and shallow rooting system of cabbage limits deep cultivation to control weeds, particularly when the soil is hard. Only when weeds are present should large plants be cultivated and all cultivation should be discontinued after heading to avoid further plant damage. The most serious weeds include tiny vetch, wild geranium, and wild turnip which must be controlled or removed prior to harvest.

Irrigation. Cabbage demands a high and consistent quantity of water for optimum development. This is attributed to its large leaf surface area and resultant loss of water through evapotranspiration. The soil water level should be maintained at 60% field capacity or higher throughout crop growth. After bolting, irrigation should be discontinued.

Harvesting, Threshing, and Drying. Generally, each parent is harvested separate from each other. However, in some cases when both parents are stongly self incompatible, they are harvested together. The uneven maturation of siliques on the cabbage raceme and the strong tendency for the seed pods to shatter creates harvesting problems for obtaining maximum yields. For these reasons, harvesting of hybrids is done by hand by cutting the seed stalks with a knife or shears. Open pollinated varieties are swathed by machine.

Conditioning. Removal of remaining chaff and straw from the cabbage seed can be done on an air-screen cleaner. Separating splits from whole seeds is accomplished with a spiral separator. In addition, cabbage seed is normally sized into several sizes for better precision planting.

Drying and Storage. Cabbage seed can remain germinable for four to six years if properly dried and stored at a low moisture content in a low relative humidity (50% or less) environment. The seed should be at a moisture content not exceeding 6% at storage. When drying is necessary, the temperature should not be above 45C (113°F) with lower moisture seeds more resilient to higher temperatures.

Lettuce (Lactuca sativa)

The exact origin of lettuce is unknown. Paintings of a cos-like lettuce are found in Egyptian tombs constructed about 4500 B.C. From Egypt, it is believed that lettuce quickly spread around the Mediterranean basin and was adopted as a popular vegetable by the Greek and Roman civilisations. The Romans further developed differing lettuce types that were broad-leaved, nonheading, nonspiny, had decreased latex content, and were resistant to early seed stalk formation. Eventually, lettuce spread throughout Europe where the present six types of lettuce were selected and produced.

Lettuce is unique among vegetable crops because it is used exclusively as a fresh, unprepared component of salads. There are six morphological types of lettuce: crisphead, butterhead, cos, leaf, stem, and Latin. Crisphead lettuce, often called iceberg, is most common in the United States. This type of lettuce produces large, heavy (up to 1.0 kg/2.2 pound) heads with tightly folded, green outer and white or yellowish inner leaves. Butterhead lettuce forms relatively small, loose heads with broad, oily, crumpled, soft-textured leaves and is commonly consumed in northern Europe.

Cos lettuce, often called romaine, has elongated leaves that are dark or light green with heavy ribs forming a loaf-shaped head and is most popular in southern Europe and the Mediterranean basin. Leaf lettuce produces a variety of leaf types and color but all are characterised by forming a rosette of leaves that do not form a head. This lettuce is used primarily in home gardens because of its ease of production. Stem lettuce is one of the few lettuce types that is not consumed raw. It produces leaves that are coarse, peeled, and cooked. This lettuce is grown primarily in the Orient.

Lettuce is relatively low in essential nutrients compared to other vegetables. It is known for its low calorie level which is attributed to a very high (94 to 95%) water content per leaf. Lettuce contains moderate amounts of vitamins A and C and calcium as well as some phosphorus, iron, sodium, and potassium. Despite its low nutritional value and lack of diversity in food preparation, lettuce is so important as a main component of salads that it ranks fourth in vegetable popularity behind only tomato, citrus, and potato.

PLANT DEVELOPMENT

Like sunflower, lettuce is a member of the Asteraceae family. The cultivated lettuce is closely related to and can easily cross with the common wild lettuce (*Lactuca serriola*). Both produce high levels of latex in the stem and leaves and this trait is responsible for the derivation of the generic name Lactuca meaning "milk-forming". Lettuce is a cool-season crop that requires abundant moisture and ample sunlight for optimum development.

Vegetative. Lettuce is an herbaceous annual that differs from its wild-type relative by producing a rosetted, short-stemmed vegetative plant. The leaves vary widely in shape from round to obovate with varying degrees of cupping and indentation. The root system develops a deep, penetrating taproot with the major absorbing surface being in the top 30 cm (12 inches) of soil.

Reproductive. Lettuce varieties are either long-day sensitive or day-neutral and high temperatures (above 18C/64°F) promote bolting (a change from the vegetative state to flower formation). In most cases, the seed stalk or inflorescence rapidly emerges from the head or leaf structure. In some cases, particularly with crisphead lettuce, the head is so tight and firm that the seed stalk is unable to emerge from the head. To enhance seed stalk emergence, three methods are used: 1) deheading - physical removal of head leaves surrounding the seed stalk, 2) slashing and quartering - cutting the seed head at its apex in an "x-shape" that loosens the tight-fitting leaves over the seed stalk, 3) application of growth regulating chemicals - application of gibberellic acid (20 to 500 ppm) to lettuce leaves at the three to five-leaf stage promotes bolting.

SEED

The lettuce fruit (seed) is an achene. It is lanceolate, flat, and the cream- to back-colored fruit wall has about 8 lengthwise ridges. The achene is 4 to 5 mm (0.16 to 0.20 inches) long, beaked, and topped with a number of large soft, capillary, white or brown, delicate pappi. The 1,000 seed weight of lettuce seed is 0.6 to 1.0 g (453,000 to 755,000 seeds per pound) dependent on cultivar.

SEED PRODUCTION

Approximately 70% of the lettuce seed in the United States is produced in California. This is due to the availability of favorable soils, climate, dependable water supply, and access to available labor since this crop requires a high labor input.

Tillage. Lettuce can grow on soils that range from fine sandy loams to clays to mucks but they must have a high organic matter content and adequate nutrients available in the upper 25 to 30 cm (10 to 12 inches) of soil. Since calcium is an important element for rapid leaf growth, the soil should be at a pH of 6.0 to 6.5 to ensure its availability. Typical vegetable tillage practices should be followed. The soil should be plowed to 30 cm (12 inches) as early in the season as possible. The seedbed is then smoothed by disking or harrowing and then leveled with a spiked harrow. When irrigation is necessary, elevated seed beds are prepared and then dragged flat.

Planting. High quality lettuce seed should be used in planting for greater uniformity of growth. It is often pelleted to facilitate precision planting. The minimum soil temperature required for germination of lettuce seed is 1.7C (35°F) and the optimum temperature is 24C (75°F). High temperatures of 33 to 35C (91 to 95°F) inhibit germination. When lettuce is planted under high temperatures, it is often planted at night and irrigated to induce germination before the high day temperatures are experienced. The seeds also can be osmoconditioned in growth regulators such as kinetin and gibberellic acid and redried, or pregerminated in a gel then planted by fluid drilling.

Fertilisation. On mineral soils, nitrogen is usually limiting and should be applied at planting, after thinning, and at bolting. Phosphorus and potassium should be applied at planting because of the crop's inability to extract these elements from deeper portions of the soil. Muck soils have a lower nitrogen requirement but require phosphorus and potassium.

Weed and Pest Control. Crop rotations with tomatoes, cucurbits, sweet corn, spinach, beets, and carrots assist in minimising the buildup of lettuce weeds and diseases. Weed control is accomplished by both cultivation and herbicide application. This practice cuts the weeds at or immediately below the soil surface, leaves a shallow mulch on the soil, and protects the small developing roots of the lettuce plant from mechanical damage. It may still be necessary for hand weeding between rows at later stages of crop development. Common weeds of lettuce fields are lambsquarter, pigweed, henbit, chicory, groundsel, sowthistle, and chickweed.

Lettuce is prone to a host of pathological disorders. From the perspective of the seed producer, lettuce mosaic virus is the most serious of these because the virus is carried in the seed. Infected seeds germinate to produce plants that are stunted, have yellow leaves, and often fail to produce a harvestable head. More importantly, sap sucking insects such as aphids feed on infected plants and serve as vectors for dissemination of the disease to healthy plants throughout the field. Seed production on infected plants can be reduced by 68% and seed weight by 62%. The virus is transmitted to 1 to 15% of the seeds maturing on the infected plant. As a result, basic seed stocks are often produced in insect-free structures or areas where temperatures are too high for aphid attack.

Most major lettuce producing areas require planting only of mosaic-indexed seed set at some local arbitrary level such as 0.5%. This level is determined by seed companies planting the seeds in flats in greenhouses and growing the seedlings to the three to four leaf stage. Infected seedlings are stunted and lighter than healthy seedlings. The use of mosaic-indexed seed has been very successful. Estimates of reductions in losses of 95 to 100% have been reported as a result of this practice

Irrigation. From seedling to harvest, lettuce demands a high quantity of water. This can be provided either by sprinklers or irrigation. The usual practice is to supply early water by sprinklers and then shift to furrow irrigation when the plant reaches the rosette stage. This allows less use of water since the sprinklers can be metered to permit moisture penetration to 30 cm (12 inch) depth per setting while the lettuce plants are small. Irrigation, however, enhances the level of water related pathogen problems. Irrigated fields should be tiled so that the water can drain through the soil thus minimising waterlogging conditions.

Irrigation for lettuce seed production is generally stopped after flowering. In particular, sprinkler irrigation after flowering causes loss of seed yield since the falling water droplets dislodge the maturing seed off of the flower receptacle.

Harvesting and Threshing. A period of 12 to 21 days is required from flowering to mature seed formation. However, because of the nature of the flowering response, not all seeds mature simultaneously on the plant. As a result, lettuce plants are harvested when approximately half of the seed heads have "feathered" (seed pappus is fully expanded and dry). They can be either machine or more commonly hand harvested.

Conditioning. Harvested lettuce seeds can be as much as one and a half times their dry weight due to contamination with high moisture flower and vegetative plant parts. These must be removed by cleaning as soon as possible to minimise seed deterioration. Lettuce seed is often cleaned on an air-screen cleaner followed by upgrading on a disk separator or an indent cylinder.

Storage. If properly stored, lettuce seeds retain their viability for four to five years. Seeds should be at a moisture content no higher than 7%. Storage deterioration of lettuce seeds is often manifested by an abnormality known as "cotyledonary necrosis". Deteriorated seeds exhibit reddish necrotic spots on the cotyledons. The cause of cotyledonary necrosis is still not known.

Carrot (Daucus carota)

Carrots originated in Afghanistan where they expanded to the eastern Mediterranean area. Arabs brought carrots with them during their conquests of Europe in the 10th to 12th centuries. European settlers brought carrots to the New World in the 1600s. Today, carrots are grown worldwide and its wild relative, Queen Anne's lace, is equally well known.

PLANT DEVELOPMENT

Carrots can be grown almost anywhere as long as the growing season is relatively cool. Their optimum growth occurs between 16 to 18C (61 to 64°F). Botanically, the carrot is an herbaceous biennial that is characterised by vegetative production in the first year of growth followed by reproduction growth after vernalisation in the second year.

Seed. The seed is fawn to brown, oblong, 3 to 4 mm (0.12 to 0.16 inches) long and 1 to 2 mm (0.04 to 0.08 inches) wide with one side flattened and the other possessing three to five ribs that are barbed. The embryo is extremely small relative to the size of the seed. The 1,000 seed weight is 0.8 g (567,000 seeds per pound).

Vegetative. Carrots form a vegetative structure consisting of an above ground rosette of leaves and a large, deep tap root. The leaves are double compound with the lower leaves of the rosette being more divided and larger than the upper leaves.

Reproductive. The carrot plant generally requires vernalisation for flowering. This effect is not related to root size since even small rooted plants, if vernalised, will bolt and flower. After vernalisation, the rosette bolts to form a stem that is 61 to 122 cm (2 to 4 feet) high with branches that produce flowering heads or umbels. The appearance of umbels is not uniform. The first and largest umbel to flower is the last umbel on the main flowering stalk and is known as the primary or king umbel. Secondary umbels are formed at the terminus of branches from the main flowering stem and flower in a sequence from the top to the bottom of the inflorescence. Tertiary umbels originate on secondary umbel stems.

Seeds produced on primary and secondary umbels vary in quality. Primary umbels produce seeds that are heavier, more mature, and of higher quality (germination percentage is higher) and vigor (germination rate is faster) than secondary umbels. As a result of this relationship, many seed companies have increased planting density in order that more primary umbels are produced for harvest. In other cases, to enhance seed yields, seed companies have established management practices that focus on harvesting the king, secondary, and tertiary umbels. This allows them to grade seeds into various sizes from large to medium to small.

Hybrid carrot seed production is possible. Male sterile lines have been discovered and in hybrid seed production the common ratio of female to male rows is 6:2 or 6:4. Unfortunately, the production of hybrid seed has required the intensive selective inbreeding of carrots for several generations resulting in inbreds with reduced plant vigor. As a consequence, yields of planted hybrid seed have not been as high as open-pollinated lines. In addition, the male sterile flowers produce smaller petals that are less attractive to pollinating bees causing lower seed set and reductions in seed yield. These factors have led to higher costs for hybrid carrot seed. This, in conjunction with lower crop yield, have continued to make open-pollinated lines the preferred choice of planting seed among carrot growers. However, the commercial fresh market acreage in the United States of the Imperator carrot varieties is almost entirely with hybrid seed. Processing carrots are still dominated by open-pollinated varieties.

SEED PRODUCTION

Carrot seed production used to be located primarily in California. However, some years prevailed when the necessary vernalisation temperatures (below 7C/45°F) to induce flowering were not experienced or high temperatures were encountered during seed set that led to empty seeds. Today, most of the carrot seed production occurs in the Columbia Basin of Washington, Madres area of Oregon as well as Idaho which are ideal desert environments for maximum seed yield and quality.

Carrot seeds are produced by two methods: seed to seed and root to seed. In the seed to seed method, the seed is planted direct in late Summer, vernalised in the Winter, flowers initiated in the following Spring, and seeds harvested in late Summer. In the root to seed method, roots raised in beds usually in an area away from commercial production are removed from the soil in the Fall, examined for trueness to type and stored at 1C (35°F) at 95% or higher relative humidity to reduce storage rots in the Winter, and then replanted in the Spring. The plant bed is used as a backup for direct seeded production in the event of Winter kill. Some types of carrots such as high sugar types do not overwinter well so they may be grown in a plant bed. In both cases, the process of removing plants, checking for off-types, storage, and then replanting is labor intensive and expensive. Even so, the root to seed method is required by most seed companies for new varieties, new inbreds and stock seed to allow selection for root characteristics. The seed to seed method, however, is preferred among seed companies for large commercial seed production.

Tillage. Fields are placed according to isolation distances established for each type of carrot. Carrots grow well on most soils but the best growth occurs on deep, friable soils such as sandy loams. Carrots tolerate a wide range of soil pH. Because seedling establishment is a problem in carrot seed production due to the seed's small size and slow germination rate, particular attention is give to tillage operations prior to planting. The seedbed must be clean of weeds, well pulverised, and compact to ensure ready flow of soil water to the seed and yet minimise puddling or crusting after a heavy rain.

Planting. If planting is done for the seed to seed method, row spacings of 50 to 90 cm (20 to 35 inches) are used and 2 to 3 kg of seed planted per hectare (1.8 to 2.7 pounds per acre). If the seed is mechanically precision planted, it is often coated using a ratio of 3:1 finished weight over the raw seed. Seeds should be planted 1.3 to 1.9 cm (0.5 to 0.7 inches) deep every 7 to 10 cm (3 to 4 inches). If the root to seed method is used, the transplants are placed every 20 to 30 cm (8 to 12 inches) apart in rows that are 75 to 90 cm (30 to 36 inches) wide.

Higher plant densities are desired in carrot seed production. Such a practice results in a concentration of umbels in the upper part of individual flower stalks causing a higher proportion of primary to secondary umbels and a more uniform period of seed maturation from umbel to umbel. Normally, bees are used for open pollinated varieties and are required for hybrid seed production. Usually 2 hives per hectare (5 hives per acre) are required.

Fertilisation. Because of the wide range of soil types in which carrots grow, a soil analysis provides the best recommendation for fertiliser requirements. As a general rule, the relative proportions of 1-2-2 are used. Less nitrogen is needed on muck soils compared to mineral soils. The nitrogen should be applied as ammonia since other forms cause serious burning. The fertilisers are best applied in a band and to the side of the seed at planting. Vegetative growth is further encouraged, particularly under irrigated conditions, by a second side dressing later in the Summer. In the Spring of the second year, another nitrogen application will stimulate vigorous plant growth prior to flowering.

Weed and Pest Control. Weed control in carrots is almost totally dependent on chemicals. This is because any physical alteration of the firm seedbed dislodges the small seed from the soil and results in poor emergence. In the past, herbicidal solvents such as Sotddbard's solvent at 702 to 749 liters per hectare (75 to 80 gallons per acre) were applied after the first true leaves were formed and before the four-leaf stage. This treatment delayed weed growth until the carrot's foliage was large enough to shade competitors. Another approach was to use shielded flame throwers to burn the weeds. Today, weed control in carrots is practiced using a combination of approved herbicides. Weed seeds that are particularly troublesome to remove during conditioning include canada thistle, nightshade, and barnyardgrass.

Irrigation. All carrot seed production is irrigated. Carrots require an abundant and well distributed water supply for optimum growth. The primary approach is to use rill (flood) irrigation. Fields irrigated overhead should be evaluated for the presence of additional fungicidal sprays. The water table should be kept at 76 to 91 cm (30 to 36 inches) below the soil surface. In inorganic soils, the water can be supplied by sprinklers and furrows. Sprinkler irrigation is most effective during the early stages of germination and emergence when the water supply does not have to be extensive. With later vegetative growth, sprinkler irrigation should be discontinued because of the increase in foliar diseases encouraged by the wet leaves.

Harvesting and Threshing. For hybrid seed production, the male rows are removed and turned into the soil prior to swathing. Harvesting of the seed crop is done by machine. This is initiated when the earliest maturing seed on the primary umbel turns brown and the umbel is brittle. Because of the uneven maturity of the seeds on the umbels and the tendency for shattering, umbels can be sprayed with a polyvinyl acetate adhesive to reduce seed loss. This process, however, is not done in large scale seed production. Harvesting early in the day while dew still remains on the umbel is also helpful. The fields are swathed by machine into windrows. After 3 to 5 days of drying, the crop is havested with modern combines at a seed moisture content of 7%.

Drying and Cleaning. The small size of the carrot seed makes it very responsive to equilibration with a low relative humidity environment. As a result, carrot seed is seldom dried unless wide ranges in seed maturity exist. Since the barbs on the seed reduce flowability and increase seed volume, they are often removed by a debearder. Cleaning of the seed is accomplished using an air-screen cleaner and indent cylinders. Carrot seed is sized into varying lengths and widths for improved precision planting.

Storage. Carrot seed should not be stored at moisture contents above 7.0%. At this moisture content and under appropriate temperature and relative humidity conditions, the seeds remain viable for 3 to 5 years.

Sugarbeet (Beta vulgaris L.)

Sugarbeet originated in the Mediterranean. It belongs to the same species as the red or gardent beet. In the middle of the 18th century, it was discovered that sugar from sugarbeet was identical to that found in sugarcane. From that point, countries reliant on sugarcane as their sole source of sugar begainefforts to develop this as an alternative sugar source. France was the first country that to breed a new variety of sugarbeet that could be economically processed for profitable yield.

PLANT DEVELOPMENT

Sugarbeet is a biennial plant that grows vegetatively the first year and produces seed the second year after being vernalised by thermal induction during the Winter. However, when grown as a sugar crop, it is harvested after the first year during which the leaves and the fleshy root are formed. In the first year, it produces a dense canopy of leaves and a large root in which the sugar is stored. In the second year, the sugar is used to form the flowering stalk, flowers, and seed. The leaves are simple with a large blade and petiole. The root is divided into three zones: The top from which the leaves are produced in a dense spiral, the thick fleshy hypocotyl where most of the sugar is found, and the root which becomes a long taproot. The flowers are perfect but incomplete; they have a five part calyx but the corolla is absent. They are produced on a panicle-like spike.

SEED

The sugarbeet "seed" that is planted is actually a single achene, which is a one-seeded fruit with a somewhat inflated, corky outer layer, or pericarp. The traditional natural sugarbeet seed unit was a multi-germ "seedball," or a cluster of two to five achenes formed by the joining of several floral units. When planted, each of the achenes germinated, resulting in three to five plants.

SEED PRODUCTION

It is important that sugarbeet seed producers have a contract with seed companies or sugar companies for seed production. This assures access to varieties that will be in demand by growers. It also assures them of parent seedstocks (i.e., inbred lines) needed to produce seed of different hybrids. Furthermore, it insures seed growers of technical advice needed for managing the planting of parental and restorer lines. Most importantly, it assures them of a market for their harvested seed.

Hybrid seed is produced in two different ways. Most commercial seed is produced by mixing the male-sterile F1 line as the primary female parent with seed of a multi-germ pollinator in a 10:1 proportion. The resulting crop contains a low percentage of seed that is not hybrid. However, since these are large and shaped differently, they are easily separated from the small flattish monogerm seed during conditioning.

The second method is more expensive, but results in a completely hybrid crop. It is used primarily to perpetuate parent stock seed which will be used in three-way or double-cross hybrids. It is produced by growing the multi-germ pollinator in about four separate rows separated by a skip-row from 16 to 20 female male-sterile rows. The male rows are destroyed after pollination and before harvesting the hybrid seed produced on the male-sterile parent.The male-sterile lines are perpetuated for further use by backcrossing with their type "O" equivalent. This process insures that the progeny will be male sterile. The type O lines are male fertile lines that do not restore pollen production to the offspring when crossed with male-sterile lines. They are comparable to maintainer lines used in producing hybrid corn.

Tillage. Sugarbeet grows best on slightly acidic fertile loamy soils. Sandy soils do not retain adequate moisture between irrigations and heavy clay soils cause problems at harvesting. Fields are plowed to 20 to 30 cm (8 to 12 inches) deep in the late Summer. If sprinklers are not used for irrigation, the fields are leveled for surface irrigation and a firm seed bed prepared by disking, harrowing, and rolling to minimise soil crusting.

Planting. Planting occurs in late August or early September. Seed is planted with a standard drill in rows spaced 51 to 60 cm (20 to 24 inches) apart at a rate of 13.4 to 16.8 kg per hectare (12 to 15 pounds per acre). Seeds are platned from 2.5 to 3.8 cm (1.0 to 1.5 inches deep. The optimum temperature for germination is 15°C (60°F).

The fields must be isolated from other fields of sugarbeets or garden beets that are allowed to produce seed. This assures genetic purity of the resulting seed. For varieties that are similar in characteristics, isolation of 0.8 km (0.5 miles) may be adequate. However, for elite seedstocks or very dissimilar varieties, isolation of 3.2 to 4.8 km (2 to 3 miles) may be desirable. It is important that the guidelines of the seed company for each variety be closely followed.

Fertilisation. Fertility practices vary greatly between seed production areas. In Oregon, the usual practice is to apply 268 kg per hectare (240 pounds per acre) of nitrogen and 112 kg per hectare (100 pounds per acre) of phosphate, along with 56 kg per hectare (50 pounds per acre) of sulfur and 5.6 kg per hectare (5 pounds per acre) of boron. Another 179 kg per hectare (160 pounds per acre) of nitrogen are side-dressed in the Spring, often in two applications.

Weed and Pest Control. Good weed control includes cultivation and chemical weed practices in addition to hand weeding. Sugarbeet grows rapidly and soon shades most weeds and prevents them from establishment. Cultivation should be used wherever possible. However, chemical control is the most common method and many chemicals are registered for all seed production areas. Finally, hand weeding is an important additional weed control practice. The most common weeds of sugarbeet seed fields are pigweed, lambsquarter, barnyardgrass, mustard, and wild oats.

Rarely can sugarbeet seed crops be grown continuously on one field. Rather, they are often rotated with cereals, beans, vegetables such as tomatoes and lettuce, and forage legumes to control diseases and root-knot nematodes. Black root is a fungal disease that kills germinating seedlings and reduces plant vigor so that uneven stands and poor yields result. Seed treatment is effective against this disease. The virus disease, curly top, is a particular problem for sugarbeet seed producers in the intermountain area. This disease is controlled almost exclusively through the use of disease resistant varieties.

Harvesting. Sugarbeet seed shatters readily as it approaches maturity. Shattering is minimised by windrowing the crop before the seed reaches full maturity and allowing it to cure a few days before combining. Windrowers are equipped with vertical cutter bars on each side to enable the crop to be cut with a minimum of shattering. In drier areas such as Arizona and Utah, this requires about 7 to 10 days compared to about 10 to 14 days in the Willamette Valley. The cutter-bar should be raised high enough so the plants are gently laid on the ground so air can circulate through the windrow and allow aeration and drying. Some of the mature seed will still shatter during windrowing, however most will fall onto the windrow and can be recovered during combining.

Conditioning. The harvested seed is first conditioned over an air-screen cleanter to remove inert material as well as multi-germ seeds which contaminate the lot. This is followed by a draper belt which further removes stems and remaining inert matter. The remaining multi-germ seed is removed by width or thickness graders which allow the small monogerm seed to fall through, but reject the large multi-germ seed. The monogerm seed are further conditioned by passing between two horizontal disks, one of which rotates. This action rubs the corky tissue from the seeds and reduces them to small, uniform size and shapes. Further screening can remove all chaff and inert material. Finally, the seeds are conditioned over a gravity table which removes light weight, low quality seed from the seedlot. The seed is then treated with a fungicide and an insecticide.

WET SEEDS

Tomato (Lycopersicon esculentum)

Tomatoes originated in Peru and Bolivia as small, wild fruits. Migration of local natives brought the tomato to Central America and Mexico where large, domesticated fruits were developed. Archaeological evidence reveals that tomatoes were a popular fruit of the Aztec and Toltec Indians who planted them with their corn. Spanish settlers to the New World carried tomatoes back to Europe where they quickly became established throughout the continent. Initially, tomatoes were grown as ornamentals because of their colorful, bright red fruits and the assumption that the fruits were poisonous.

PLANT DEVELOPMENT

The tomato is a warm-season herbaceous perennial that requires three to four months from seeding to production of fruit. It grows best at temperatures between 18 and 29°C (65 and 85F), is often killed at temperatures below freezing, and does not grow at temperatures above 35°C (95F).

Seed. Tomato seeds are 3 to 5 mm (0.12 to 0.20 inches) in size, flat, reniform, and have a grey, slightly hairy seed coat. The embryo is large and coiled in a limited, fleshy endosperm.

Vegetative. Tomato plants are typically viny, prostrate, and are either determinate or indeterminate based on whether the apical stem terminates in an inflorescence. Most shoots form in the axils of leaves. The tomato plant produces a deep tap root with extensive secondary roots.

Reproductive. Tomato flowers are yellow and formed on racemes. The stamens are fused with the lower part of the corolla forming a staminal cone. Generally, pollen release occurs prior to elongation of the style through the staminal cone resulting in self-pollination. However, temperature has a major effect on this process. If temperatures are below 15°C (60F) or above 29°C (85F), pollen release is restricted resulting in incomplete fertilisation of ovules. This causes collapsed fruit walls and formation of deep indentations in the fruit, a phenomenon called "catface." Hot, drying winds also cause tomato flowers to abscise from the plant. The best shaped and fullest tomato fruits result from optimum fertilisation of ovules. The mature tomato fruit is botanically a berry. Fruit color varies from the common red to yellow, pink, and orange.

SEED PRODUCTION

Seed production in greenhouses has occurred on a limited scale. As a result, most of the following discussion will consider tomato seed production as practiced under field conditions.

Tillage. Crop rotations are useful in tomato seed production because of the crop's susceptibility to a variety of diseases. Rotations with soybeans, sugar beets, wheat, and beans have been successful. One of the most common rotations in the California San Joaquin Valley is with cotton. Fields used in tomato seed production should be Fall plowed to a depth of 20 cm (8 inches). This practice not only improves the physical structure of the soil but also encourages the decay of roots and other plant parts thereby enhancing the level of soil organic matter. If organic matter remains low, manure can be applied at a rate of 13,440 to 17,920 kg per hectare (6 to 8 tons per acre) and incorporated into the soil at the time of Fall plowing.

Planting. Tomatoes can be planted either by direct seeding or transplants. When 152 cm (5 foot) rows are used, seed is planted to give a spacing of 20 to 30 cm (8 to 12 inches) or 30 to 40 cm (12 to 16 inches) if double rows are used. Planting should not begin until the soil temperature has reached 10C (50°F) since germination will not occur below this temperature. The optimum range of soil temperatures for seed germination is between 15C and 29C (60°F to 85°F).

Direct seeding is accomplished by precision planters that equally space the desired number of seeds in the row. Seeds should be planted at a depth of 1.3 to 2.0 cm (1/2 to 3/4 inch) to obtain a population of 24,700 to 49,500 plants per hectare (10,000 to 20,000 plants per acre) in single rows. Planters should be modified to allow the application of anticrustants (keep the soil from crusting and thereby facilitating uniform seedling emergence) and starter fertilisers applied either in a solid or liquid form.

Transplanting is another important planting approach used in tomato seed production. Transplants can be obtained either from local greenhouses or can be from transplants produced using grower seed. They are at the correct stage for planting when the first pair of seedling leaves appear. When possible, transplanting should be done either during the afternoon or on a still, cloudy day to avoid transplant shock. The transplants should be thoroughly watered before planting and a starter fertiliser high in phosphorus added. Transplants every 40 to 45 cm (16 to 18 inches) in rows 168 cm (66 inches) apart provide a population of 14,800 to 19,700 plants per hectare (6,000 to 8,000 plants per acre).

Fertilisation. While tomatoes grow on almost any soil, soils that are sandy, light, well-drained, and high in organic matter and mineral nutrients produce the best quality and highest fruit yields. Optimum soil pH should range from 6.0 to 6.5. Tomatoes require an abundance of the three major elements: nitrogen, phosphorus, and potassium. Adequate nitrogen is important to enhance foliage growth which has a major bearing on crop maturity and protects fruits from sunscald. Phosphorus influences fruit quality by stimulating vigorous root growth that enables more nutrients to enter the plant thereby promoting sturdy stem growth and healthy leaf formation. Tomatoes use large amounts of potassium. This element is important in stimulating early plant growth and regulating normal carbohydrate and protein metabolism.

Fertilisers are generally applied broadcast before Fall plowing. All the phosphorus and potassium and one-half the nitrogen are added at this time. Between the time the plants have two to three leaves and when the flowers are first observed, a side-dress nitrogen application is useful. Nitrogen should not be applied after fruit formation because it stimulates vegetative growth.

Weed and Pest Control. Mechanical and chemical control of weeds is practiced in tomatoes. Cultivation to eradicate weeds begins soon after seedling emergence. Early cultivations are close to the seedling and 2.5 to 5.0 cm (1 to 2 inches) deep. Later cultivations are further from the plants and shallower to minimise injury to rapidly spreading surface roots. Chemical control also involves application of an array of pesticides specific to the weed problem. Weeds that are difficult to control in tomato fields include nightshade, pigweed, lambsquarters, green foxtail, wild oats, nutsedge, smartweed, and morningglory.

Harvesting. Tomato fruits can be hand or machine harvested. In the case of hybrid seed production, the marked or tagged fruits are always hand harvested. All mechanical tomato harvesters are based on a "once-over" principle in which the entire tomato plant is harvested. They have four components: pick-up mechanism, fruit and vine separating area, fruit sorting area, and container loading mechanism.

REFERENCES

Allard, R. W. 2020. *Principles of Plant Breeding*. John Wiley and Sons, Inc. New York

Chin, H. F. 2018. "Storage and Vigour". *Seed Science and Technology* 16:1-4.

Douglas, J. E. ed. 2021. *Successful seed programs: A planning and Management Guide*. International Agricultural Development Service. Westview Press Boulder, Colorado.

IRRI. 2018. Rice Seed Health. Proceedings of the International Workshop on Rice Seed Health, 16-20 March 1987.

Mew, T.W. and Misra, J.K. 2014. *A manual of rice seed health testing*. International Rice research Institute. Los Banos, Laguna, Philippines. P.O. Box 933, 1099 Manila, Philippines.

12

Seed Storage

A major part of the human diet all over the world consists of cereals and legumes. According to FAO estimate, 70% of human food comprises cereals and legumes and the remaining 30% comes from animals, which again feed on these seed meals. Animal proteins being more expensive, people in developing countries virtually depend on seed protein alone for their entire protein requirement. But unlike animal proteins like casein and egg albumin which are nutritionally more balanced in terms of all essential amino acids, plant proteins are generally deficient in some of them. As human beings cannot synthesise essential amino acids, these must be supplied from outside as food. Cereal proteins are generally deficient in lysine and tryptophan whereas legumes are deficient in sulphurcontaining amino acids—methionine and cysteine. The presence of antinutritional compounds like glucosinolates (thioglucosides) in *Brassica* and proteinaceous toxic compounds like lectins, enzyme inhibitors, etc. in legumes adversely affects the nutritional quality of seed proteins. Thus higher or better quality with reference to seed proteins can have many different meanings, depending upon the crop and its utilisation. But by convention, the 'quality' of seed protein refers to how far its amino acid composition matches the balanced amino acid composition recommended for human diet by the World Health Organisation (WHO).

Before the advent of recombinant DNA technology, studies on seeds mainly focused on the isolation, characterisation and nutritive value of proteins they contained. In the last fifteen years, however, the ability to isolate, manipulate and reintroduce genes into other plant species has led to an enhanced interest in the means for qualitative and quantitative improvement of seed grains by genetic engineering.

The best way to store seeds is to package them in paper envelopes or bags since they allow for good air circulation and don't sweat. However, any container will do, keeping in mind that humidity and lack of air circulation will cause mold, disease and prompt seeds to germinate prematurely. Film canisters for one aren't recommended as the plastic promotes humidity and stagnant air. The temperature should be cool to make longer storage possible-refrigerator storage will work if you can't find a naturally cool place. Be sure to write the date, name of plant and any growing instructions you are aware of on the envelope or package. This will come in handy when using the seeds a year or more later, and will be appreciated if you give the seeds to someone else. It's worth it (but not necessary) to put a bit of extra effort into the packaging if the seeds are to be given away. Use specialty papers for the envelopes or create fancy labels to mark them. Some envelope and label templates are provided here for you to print out on any paper (or sticker paper for labels that is compatible with your printer type).

Store seeds carefully by placing envelopes inside large glass jars with a bag of silica or powdered milk. These products absorb excess moisture. Reuse the tiny bags of silica gel that come inside new shoes—dry them for a few minutes at a very low temperature in your oven. Alternatively, make a tiny package of powdered milk by pouring a pile into the center of a piece of breathable fabric or tissue paper. Pull the corners together and close it up with a piece of string or elastic to create a sachet. The best jars for storage are wide mouth mason jars used for canning. They have the proper airtight seal that is essential for long term storage. If you store the jars in a cool, dark place the seeds should last from a year to a few years, depending on the type.

FACTORS AFFECTING SEED STORAGE

Conditions essential to good seed storage are just the opposite of those required for good germination. Good germination occurs when water and oxygen are present at a favorable temperature. Good seed storage results when seeds are kept dry (below 8 percent moisture) and the temperature is kept low (below 40 degrees).

When seed moisture and storage temperature are low, the presence of oxygen has not been shown to be a factor in seed longevity. Germination is unaffected by storage in atmospheres of nitrogen, carbon dioxide, partial vacuum or air.

Relative humidity (RH) influences the moisture content of seed if it is not stored in moisture-proof containers. For example, at 15 percent RH, seed will dry down to 6 percent moisture and will store safely in this condition for several years. However, at 90 percent RH, seed will dry down to only 19 percent moisture and germination will be poor after one year.

The drier the seeds, the longer they will store. There is a chance of producing what is known as "hard seed" if moisture is reduced below 8 percent. Hard seed resists germination under favorable conditions because it does not absorb enough water. When planted, the seed gradually absorbs water, germinates and produces seedlings over an extended period. A seed lot containing 50 percent hard seed is little better than a lot containing 50 percent dead seed, because neither produces a stand of seedlings when they should. Beans and peas are particularly subject to this condition and therefore should not be dried as completely as other seed. If they have been overdried, they germinate better if exposed to a humid atmosphere for two weeks before planting.

ISOLATION OF SEED PROTEINS

Seed proteins can be broadly classified into two categories, viz. housekeeping proteins and storage proteins. The housekeeping proteins are responsible for maintaining normal cell metabolism. A more recent classification divides these proteins into storage, structural and biologically active proteins. The major biologically active proteins include lectins, enzymes and enzyme inhibitors. These are minor proteins and may have nutritionally more balanced amino acid composition than storage proteins. The seed storage proteins on the other hand, are nonenzymatic and have the sole purpose of providing proteins (nitrogen and sulphur source) required during germination and establishment of a new plant. Seed proteins were empirically classified many years ago by Osborne based on their solubility as follows:

Albumins: Water extractable (1.6S–2S);

Globulins: Extractable in dilute salt solutions (7S–13S);

Prolamins: Extractable in aqueous alcohol;

Glutelins: Most difficult to solubilize; Extractable by weakly acidic or alkaline or dilute SDS solution.

Albumins and globulins comprise the storage proteins of dicots (e.g. pulses), whereas prolamins and glutelins are major proteins in monocots (e.g. cereals). The seed storage proteins can be distinguished from other proteins by some of their characteristics, e.g. (1) These accumulate in high amounts in seed during mid-maturation stage of seed development and are used up during germination. (2) These are synthesised only in the seed (in cotyledon or in endosperm) and not in other tissues. (3) They lack any other functional activity besides storage. (4) They are deposited mostly in special storage organelles called protein bodies.

Isolation

Seed proteins are generally isolated by extracting crushed whole seeds or seed meals left after oil extraction (defatted by solvent extraction, if necessary) with a low salt buffer. When desired, the different proteins are partially separated by successive differential extraction with water, salt solution, aqueous ethanol and then acid/alkali with detergent. The total or partially fractionated proteins are further purified by a combination of methods like heat and isoelectric precipitation, ammonium sulphate fractionation gel filtration and ion-exchange chromatography. For further separation of microheterogeneous forms/types, techniques like isoelectric focusing, two-dimensional electrophoresis and HPLC are used.

As expected of a reserve of nitrogen for the seedling, storage proteins are generally rich in asparagine, glutamine and arginine or proline. The legume proteins (mostly 7–11S globulins) tend to be deficient in the sulphurcontaining amino acid cysteine and methionine and also tryptophan while cereals generally lack lysine, threonine and tryptophan. The globulins are generally present as oligomers. The cereal prolamins are present as monomers or small aggregates, while glutelins form large disulphidebonded aggregates. Most of the seed proteins have high molecular weights and their water solubility is poor. There are many examples of small water soluble seed proteins in dicot seeds like castor bean, pea, mung bean, mustard, *Amaranthus, Chenopodium,* Alfalfa, many oil seeds, etc.

Storage proteins have also been studied from many seed grains of importance in India like chickpea, pigeon pea, black gram, mung bean and several oilseeds.

All these studies revealed a close similarity in sizes, subunit nature and other physico-chemical properties of the same seed protein fraction from different legumes. For example, Uppal and Matta working with black gram or urad bean (*Vigna mungo*) and Krishna and Bhatia with pigeon pea (*Cajanus cajan*) found the major globulin fraction as 7S vicilin, a trimeric protein composed of one large (70–72 kDa) and two small (40–57 kDa) subunits not linked by disulphide bridges. This contrasts with the major 7S vicilin from pea which is a homotrimeric protein of 70 kDa subunits, also not linked by disulphide bridges. These proteins are normally heterogeneous in terms of polypeptide size, charge and amino acid composition. Pea legumin or vicilin, a comparatively simple storage protein, contains up to 30 separable polypeptides as observed in two-dimensional isoelectric focusing/ SDS-PAGE analysis, whereas a complex fraction such as wheat storage protein contains more than 60 polypeptides. In oilseed *Brassica* and other cruciferae, seed storage proteins are 2S albumin (napin) and 12S globulin (cruciferin) types. The 2S albumins are synthesised as large precursor molecules and after post-translational processing a signal peptide, an additional amino terminal fragment, an internal fragment and some carboxy-terminal residues are removed. The two subunits resulting from this cleavage process are linked by disulphide bridges. Similar is the case with 2S albumin from *Arabidopsis thaliana* for which at least five isoforms have been characterised. Castor bean 2S protein is unique in that two heterodimeric 11 kDa proteins of 97 and 99 amino acid residues are formed by processing of a single polypeptide synthesised by a single mRNA. Eight members of the 2S albumin of sunflower are made of a single polypeptide chain, of which SFA 8 is rich in sulphur amino acids (16 Met and 8 Cys). In addition, Allen *et al.* reported the deduced amino acid sequence of another napin-like sunflower 2S albumin, HaG 5, encoding heterodimeric protein with approximate molecular weight of 14 kDa. The 12S globulins are oligomeric complexes consisting of six subunit pairs, each subunit being composed of a heavy alpha chain and a light beta chain. Each pair is derived from a single precursor. In the 12S globulins of legumes, cucurbits, rape seed and oat, there are oligomeric structures of six subunits, in which each subunit is composed of disulphide-bonded large acidic and small basic polypeptides. The mechanisms of synthesis are also similar in different genera. Similar subunit nature of 2S albumins and 12S globulins from *Chenopodium album* and *Sterculia foetida* has also been described.

Amino Acid Composition

A large number of purified proteins have been characterised in terms of their amino acid composition for their importance in human and animal nutrition. In general, it has been found that cereal and oilseed proteins are nutritionally more balanced in amino acid composition compared to legume seed proteins. Table 1 shows the amino acid composition of some representative seed proteins compared with the WHO recommendation of essential amino acid content. A comprehensive early information on the isolation and composition of various seed proteins can be obtained in Prakash and Rao. It can be seen that some mustard, *Chenopodium* and *Amaranthus* seed proteins are almost ideal proteins in terms of essential amino acids and meet or surpass WHO recommendation. The same holds true for 2S protein from cotton seeds and buck wheat. Also, most of the Asp and Glu residues in seed proteins are present as Asn and Gln, as was determined from nucleotide sequences later.

Understanding the structural features of seed storage protein is important for providing a basis for proposed engineered mutations at suitable sites of these proteins with the ultimate goal of enhancing nutritional quality, without affecting their functional role in seed metabolism and storage. In spite of large complexity in polypeptide subunits and amino acid sequences, diverse seed proteins may share considerable structural and sequence homology. This only indicates that notwithstanding evolutionary divergence, seed proteins have retained some essential structural features due to some functional constraint in seed metabolism and/or their function as storage proteins in specified organelles. The 3D structure of seed proteins has been studied by spectroscopy, X-ray diffraction and other physicochemical physicochemical techniques. One general feature of the proteins is their more or less compact structure with a hydrophobic core and some exposed loops or surface domains.

Table 1. *Amino acid composition of some seed proteins (mole%) compared with WHO recommendation of essential amino acids*

Amino acid	WHO	Maize zein	*B. camp.* (yellow sarson) 2S†	Chenopodium (goose foot) 2S protein	Amaranthus AmA†	Pea legumin
Ala (A)	—	13.7	4.2	4.3	5.3	6.0
Arg (R)	—	1.2	4.2	7.5	5.3	10.0
Asx (B)	—	4.7	1.6	5.0	16.6	13.1
Cys (C)	—	Trace	6.6	5.8	0.7	1.2
Glx (Z)	—	21.2	23.9	17.9	8.9	19.7
Gly (G)	—	2.1	6.6	5.8	12.0	6.9
His (H)	—	0.9	3.3	5.7	3.3	1.8
Ile (I)	4.0	4.1	4.2	3.8	5.0	4.0
Leu (L)	7.0	20.0	6.6	3.0	7.6	7.6
Lys (K)	5.5	0.2	6.6	4.9	6.6	4.2
Met (M)	3.5*	Trace	3.3	2.8	1.6	0.7
Phe (F)	6.0**	6.3	1.6	5.4	5.6	2.3
Pro (P)	—	9.7	8.2	6.1	3.6	5.5
Ser (S)	—	6.6	4.9	5.6	6.6	5.8
Thr (T)	4.0	2.4	3.3	5.1	5.3	3.1
Trp (W)	1.0	N.D.	0.82	N.D.	2.3	—
Tyr (Y)	**	3.1	0.82	3.0	4.8	3.6
Val (V)	5.0	3.6	4.9	3.4	5.6	5.0

†, Derived from DNA sequence; N.D., not determined; *, Met plus Cys; **, Phe plus Tyr.

Published reports on direct structure determination of seed storage proteins by X-ray diffraction are scanty. Some protease inhibitors (many of which have amino acid sequence homology with seed storage proteins) have been well characterised both by X-ray diffraction and by NMR studies. De-Meester *et al.* determined the 3D structure of soybean Kunitz-type trypsin inhibitor by molecular replacement using the coordinates of homologous *Erythrima* trypsin inhibitor. It was found that the conformation of the reactive site loop did not appreciably change by interaction with trypsin. The structure of a seed globulin having chymotrypsin inhibitory activity from winged bean *Psophocarpus tetragonolobus* has been determined by X-ray crystallography. The protein structure revealed 12 antiparallel beta-strands with connecting loops arranged in a characteristic beta-trefoil fold common to other homologous serine protease inhibitors. The X-ray structure models of proCon A and Con A are available. This study confirmed that the residues involved in transpeptisation reaction during the formation of mature Con A are neighbours. Sainani *et al.* studied the effect of chemical modification of nine different amino acids separately in pennisatin, a major storage protein of pearl millet, on the structural and functional properties. As expected, changes in hydrodynamic properties indicated a change in the secondary structure for most substitutions. Thermostability remained unaffected except in the case of Glu and Asp modification. Some essential conserved features in the primary structure of napin-like 2S albumins have been described. By proteolytic cleavage and subsequent peptide mapping by matrix-assisted laser desorp-tion ionization mass spectroscopy (MALDI-MS), the correct assignment of 8 cysteine residues (2 in the small and 6 in the large subunit) of napin involved in 4 disulphide linkages has been possible. The high thermal stability of napin tertiary structure was corroborated by CD and high sensitivity differential scanning calorimetry. Two consecutive transitions were attributed to two different folding domains.

Limited proteolysis with several endopeptidases, immunological comparison and CD spectrometry indicated that cleavable propeptide regions are exposed to the exterior of the mature napin. Moreover, the conformation of the mature napin was not significantly different from the pronapin expressed in insect cells. Chemical, CD and fluorescence spectroscopic studies with the isolated small subunit of napin indicated that it has a strong tendency to form a non-native intramolecular disulphide bridge. Addition of the reduced large napin subunit inhibited this tendency. This study indicates that interchain disulphide formation in napin may be an early step during maturation of this (and other) multisubunit seed protein. Using information from CD and NMR spectroscopy and sequence homology, Pandya *et al.* proposed a model for the structure of sunflower 2S albumin SFA8, based on the known structure of ragi alfa-amylase/trypsin inhibitor. The model is composed of a bundle of four alfa-helices tightly knit by 4 disulphide bridges. From CD and chemical denaturation studies, we proposed a similar structure for *Brassica* 2S albumin (Ghose *et al.* unpublished observation). From thermal denaturation, we observed no intermediate molten globule state for this protein.

Apart from the structure, other physicochemical and hydrodynamic properties of seed proteins are important determinants for their use in the food industry, e.g. viscosity, thermal-pH-emulsion stability, elastoextensibility, baking cooking quality, etc. Numerous studies have been carried out with different oilseeds, legumes and cereals.

Homologies of Seed Storage Proteins

With the increasing knowledge about the sequence of different proteins and efficient computer programs, homologies between drastically unrelated proteins are being detected. The homology studies not only give an insight into the structural and functional properties of different proteins but more importantly cast light on the evolutionary status of these proteins. The homologies of seed proteins can be studied by comparing their cross reactivity with specific antibody against one and amino acid sequence, either directly determined from protein or more easily derived from nucleotide sequence of cloned cDNA. The different classes of seed proteins exhibit considerable homologies among themselves. Several diverse 2S proteins from different dicot seeds have shown sequence homology: the methionine-rich brazil nut seed protein, the allergenic seed protein in castor bean, the very basic 2S albumin in rape seed, a trypsin inhibitor from barley and albumins of sunflower and wheat. The amino acid sequences of the coding region of 2S storage albumin of *Brassica* sp. showed 80% identity amongst themselves.

Homologies between seed proteins have also been studied with the help of antibody cross-reactivity studies and subsequently finding the homologous domains. Antigenically homologous seed proteins were found in different phylogenetically related species of *Cruciferae* and of Chenopodiaceae and Amaranthaceae family. Common antigenic determinants were in the prolamin fractions of wheat, barley and maize. N-terminal sequence analysis of wheat and barley prolamin shows homologous regions. Prolamin c-DNAs bear strong homologies, 11S globulin cDNA of oat shares 74% homology with rice glutelins. It has been shown that parts of the non-repetitive domains A, B and C of one group of prolamins from barley, rye and wheat have considerable sequence similarity with sequences present in a large group of seed proteins from monocot and dicot plants, including Bowman–Birk protease inhibitors, cereal inhibitors of alfa amylase and trypsin, and 2S albumin storage proteins of castor bean, sunflower and oil-seed rape. This implies an ancient origin of these domains from prolamin super family whose overall topology has been conserved in spite of large allowable changes in amino acid residues.

Storage proteins are encoded by families of polymorphic genes and gene copy number can be very high. The *Arabidopsis thaliana* family for example, contains four tandemly clustered members. The globulins of dicot seeds are also the products of multigene families. In monocots, one well-studied seed storage protein is zein of maize. It has been suggested that there may as many as ten families of prolamin in maize, each having 3 to 10 genes. This means that there are 30 to 100 zein genes in maize. In spite of microheterogeneity in zein families, there is 90% homology between the nucleotide sequences. The microheterogeneity gives rise to mostly conservative mutations, so that in the amino acid level the replacement is mostly by homologous amino acid which does not alter the overall structure and solubility of the protein. In a class of zein genes, there are 6 to 9 repetitive blocks. This type of block structures are also found in wheat prolamin and gliadin. It has been suggested that homology which extends over such diverse plants, both monocots and dicots, must be due to constrained protein structure necessary to fulfil the basic physiological rôle of seed proteins as a nitrogen source for developing seedlings. Multigene families of seed proteins indicate that probably they are evolved by a complicated series of gene duplication. It could be implied that some non-reciprocal recombinational events occurred in the evolution of these genes.

Disposing diverse seed protein gene variants can be advantageous for the plants in some cases. Shewry *et al.* have proposed a hypothetical evolutionary scheme of multigene families based on extensive sequence comparison. Both gene duplication and DNA exchange might have played a role in the evolution of 2S albumin genes of Brassicaceae.

Synthesis, Processing and Deposition of Seed Storage Proteins

Proteins comprise one of the major and important storage materials which accumulate in high amounts during the second stage of seed development, i.e. the mid-maturation stage. This stage is preceded by the zygote development stage and followed by the desiccation stage in which seeds undergo preparation for dormancy. During the first phase, cell division within the zygote gives rise to a globular embryo which differentiates into the embryonic axis and the cotyledons. The fate of the endosperm and cotyledon at this stage is different in monocots and dicots; in monocots the endosperm becomes the major storage tissue while in most dicots the endosperm has only a transient presence and the cotyledons fulfil the major storage function.

During the second phase, cell expansion takes place, the embryo enlarges and there is rapid synthesis and accumulation of food reserves. In the pea seeds, the protein is stored in the protein bodies whereas starch is stored in plastids of cotyledon cells. In general, there is a sudden increase in storage protein which occurs within a few days during this period. Monocot plant species have specialised storage tissue, the endosperm (triploid) or perisperm (diploid) where storage reserves are laid down. This is followed by the third stage which consists of developmental arrest and desiccation which prepares the seed for dormancy. The seed loses more than 90% of its water content, RNA and protein synthesis terminate and embryonic dormancy begins. Although metabolic processes and *de novo* protein synthesis drops dramatically during this stage, late embryogenesis associated (LEA) proteins, which may be stress proteins produced to protect the seed tissue against desiccation, are synthesised at this stage. Dormancy ends in seed germination when seed storage proteins are used up for the seedling development. The duration of different stages of seed development varies with the species and environmental factors.

Seed proteins are usually synthesised as precursors having a leader pre-peptide sequence and pro-sequence which contain the signal for transport, processing and targeting from the site of synthesis to the storage organelles. The signal sequence is instrumental in directing the nascent chain from the rough side of the ER to the ER lumen. It is now well accepted that all storage proteins are synthesised on the rough ER. After cotranslational cleavage of the signal peptide, the polypeptides are subjected to glycosylation in specific cases and protein disulphide isomerase-catalysed disulphide formation, folded with the help of chaperones and targeted to protein bodies. Protein bodies are formed either by budding-off of storage proteins from ER (as in the case of zein and rice prolamin) or from protein storage vacuoles via the golgi apparatus. A more detailed discussion on processing and deposition of storage protein is presented in a recent review by Muntz.

Most seed storage proteins undergo limited proteolytic processing after initial removal of the signal peptide. The exact subcellular site of this processing and the enzymes involved may vary for different seed proteins and have not been fully elucidated yet. Some excellent reports, however, have appeared in recent years. Muntz and his group made a series of interesting studies on the processing, disulphide bridge formation and targeting of 11S globulins. From *in vivo* (in transgenic tobacco seeds) and *in vitro* transcription/translation experiments, they showed that disulphide bridge formation probably occurs co-translationally and is necessary before the processed chains can form multimeric structures. Prolegumins are thought to be processed into alfa and beta chains in the protein storage vacuoles by a cysteine endopeptidase (cysteine in the active site) with Asn or Asp cleavage specificity. The napin-like 2S albumins of rapeseed, mustard and Brazil nut show more complicated processing. After cotranslational cleavage of the signal peptide in the ER, they undergo limited proteolytic processing in the vacuole whereby a propeptide, a short linker peptide between the small and large subunits of the mature protein and a short C-terminal peptide are cleaved off. Controversy exists regarding the processing enzymes and recognised amino acid sequence motifs. The propeptide cleavage sites have an Asn (or Asp) at or near the p1 position and the detached propeptides are degraded by aspartate protease(s), previously indicated to be the processing enzymes. Asparaginyl endopeptidase (with cysteine at the active site) has strict substrate specificity towards carboxy side end of asparagine residues. A *Vigna mungo* processing enzyme termed VmPE-1 of this type is indicated to be involved in the vacuolar processing of yet another cysteine endopeptidase SH-EP which degrades seed storage proteins during germination.

From studies on 2S albumin of *Arabidopsis*, Krebbers *et al.* indicated the occurrence of a Gly–Pro motif as a prerequisite for proteolytic processing, but subsequent sequence comparisons with *Brassica* napins and Brazil nut protein questioned the universality of this motif. It is speculated that some generally shared conformational features rather than amino acid sequences surrounding the cleavage sites are required for such processing. Yamada *et al.* recently reported an interesting case of the formation of multiple functional proteins by cleavage of Asn-Glu bonds of a single precursor by vacuolar processing enzymes in maturing pumpkin seeds. The processing of proconcanavalin A (pro Con A) into Con A needs special mention here. Mature Con A from jack bean is a homotetrameric protein of unglycosylated

30.4 kDa subunits of 237 amino acids. After linker peptide deglycosylation, there occurs a circular permutation. In other words, after excision of internal 15 residues linker peptide, a head to tail transpeptidation (protein ligation) takes place which joins the N-terminal Ser of pro Con A via a peptide linkage to an Asn site, 9 residues upstream from the C-terminal of procon A. The C-terminal nona peptide is simultaneously detached.

Regulation of Storage Protein Synthesis

The developing seed provides an attractive biological system for study of highly regulated gene cascade. Of the large number (about 20,000) of diverse genes expressed during embryogenesis, most encode rare mRNAs of unknown function. A small gene set, however, directs the synthesis of abundant mRNAs that encode seed storage proteins that are packed preferentially into protein bodies in the embryo, dicot cotyledons and monocot endosperms. A large number of seed protein genes have been characterised at the molecular level. These genes are excellent models for dissecting the different spatial, temporal and environmental regulatory processes. A number of reviews partly deal with some aspects of seed protein regulation.

Spatial Regulation

There are several reports indicating qualitative as well as quantitative differences in the accumulation of seed storage proteins within different seed tissues. In cereals such as wheat, barley and maize, the major storage protein, prolamin, is deposited only in the starchy endosperm, whereas albumins and globulins tend to be largely concentrated in the embryo and aleurone tissue. Although albumins and globulins are found throughout the legume seed, the outer cell layers of the abaxial cotyledon surfaces of the French bean and cowpea have been shown to contain higher levels of albumin protein. In addition to storage proteins, several other proteins like lectins, proteinase/amylase inhibitors, LEA proteins are also synthesised in the embryo. Apart from possibly serving as storage proteins, these proteins have also been implicated as defence proteins against several biotic and abiotic stresses. Such differential spatial (tissuespecific) accumulation of proteins may provide important clues to the function and regulation of these genes.

Developmental Regulation

There are temporal controls of seed protein gene expression, as shown by numerous reports on a variety of seeds. This has been evident by detailed analysis of seed protein accumulation, measuring mRNA levels by cell-free translation and by nucleic acid hybridisation using labelled cDNA sequences specific for each mRNA or a family of mRNA. These studies have revealed that each storage protein mRNA has a characteristic period during which it reaches a maximum level and then declines, although they all overlap. For example, vicilin synthesis precedes legumin synthesis in *Vicia faba*. In general, the proportion of low molecular weight pro-teins decreases along with concomitant increase in high molecular weight proteins as the seed matures. The low molecular weight proteins have a better amino acid composition. As a result, green seeds are nutritionally better than ripe ones. Presumably, genes in each set respond to distinct regulatory signals, but the actual molecular interaction and mechanism of regulation could be effected at two levels. The short range control involves the sequences immediately upstream and downstream from 5¢ and 3¢ ends of the genes. The long range controls, although not identified like the former, may involve remote DNA sequences, genes or other structural features within chromosomes. Such understanding is essential for practical application of genetic engineering tools for the improvement of protein quality and/or quantity.

Genetic Regulation

Studies on the rates of accumulation and final levels of storage proteins in different genotypes and mutants of a plant provided clear evidence of genetic regulation of these proteins. In some cases, the defect is due to a mutation in the seed protein gene itself; while in others, the defect is due to a mutation in unlinked regulatory sequences functioning as transacting loci. The seed protein gene mutations are caused by several different events including deletion of an entire locus, insertional inactivation, point mutations that alter the protein reading frame, deletion of a 5¢ flanking region and chromosomal rearrangement. The mutations either eliminate or greatly reduce the levels of seed proteins or their mRNAs. For example, the mutant soybean lectin *le-1* gene is not transcribed detectably as a result of an insertion element present in the gene itself. In maize, the opaque-2 mutation causes a reduction in the transcription of many unlinked zein storage protein genes. Pysh *et al.* demonstrated that *opaque-2* gene encodes a polypeptide belonging to the basic domain/leucine zipper class of transcription regulatory proteins and controls the expression of the 22 kDa zein proteins by binding to its promoter region. A defect in the seed protein gene may also be reflected in the final level of protein through its effect on the posttranslational processing and targeting. Cotranslational cleavage of the signal prepeptide and removal of the prosequence of seed protein in the lumen of ER are necessary necessary for correct processing, folding and targeting to vacuolar protein bodies. For example, fleury 2 mutant of maize having a mutant alfa-zein protein deficient in signal peptide cleavage has a much reduced level of alfa-zein and defective protein bodies. Different regions of the N-terminal and C-terminal and even internal seed protein sequences have been shown to be involved in their correct targeting and deposition.

Environmental Regulation

Environmental factors such as temperature, nutrients and hormones may alter seed protein composition and quantity. Seed protein composition is altered by temperature in wheat and pea. Nutrient deficiency particularly of sulphur, selectively and markedly depresses sulphur-rich proteins compared to other seed proteins in rape, legumes and cereals. A high osmotic pressure in tissue culture medium initiates legumin synthesis in cultured immature pea cotyledons and 12S globulin in rapeseed. Abscisic acid has also been shown to stimulate storage protein synthesis in cultured cotyledons of soybean and rapeseed. It is assumed that the same conditions may regulate storage protein *in vivo* also.

Regulatory Sequences in Seed Storage Protein Genes

The spatial and temporal controls of seed protein gene expression are exerted at the transcriptional level. They act like switch processes which may be turned 'on' and 'off' in the right tissue at the right time. Both *cis*-acting elements at the 5¢ flanking regions of the protein genes and *trans*-acting binding protein factors are involved in this switching mechanism. Some conserved nucleotide sequences have been found in many well-characterised storage protein genes. Apart from 'TATA' and 'CAAT' boxes, some important regulatory elements have been identified by gel mobility shift on protein binding, and expression of a reporter gene product like CAT (chloramphenicol acetyl transferase) or GUS (beta-glucuronidase) linked to deleted promoters in transgenic plants. In the 11S legumin-like genes from pea, soybean, broad bean (*Vicia faba*) and sunflower, a conserved sequence of approximately 28 bp called the 'legumin box' (5¢-TCCATAGCCATG-CATGCTGAAGAATGTC-3¢) is located within 250 bp of the transcription start site (TCS). The legumin box sequence has been implicated in the regulation of expression of alfa-subunit of betaconglycinin in soybean at the transcriptional level. Another sequence termed the 'Vicilin box' (5¢-GCCCACCTCA/TTTTC/TGTTC/TAC/TTTCAACACNC-GTCAANNTNC AT-3¢) located within 150 bp of the TCS in the 5¢ region has been found in 7S protein genes in pea, soybean and *Phaseolus vulgaris.*

In the cereal prolamin genes of barley, wheat and maize, a sequence 300 bp upstream of TCS, termed the '-300 box' has been found. The ATGAGTCA motif in the -300 element of rice glutelin gene has been found to be involved in nuclear protein binding. The low mol. wt. glutelin gene *lmwg-1D1* promoter in wheat contains the 'endosperm box', a sequence motif highly conserved in the promoter region of a large number of monocot storage protein genes and is thought to confer endosperm-specific expression of prolamin genes.

A class of elements of alternating purine–pyrimidine sequence called the RY repeats (5¢-CATGCATGT-3¢) is located in multiple copies in most legume seed protein genes and also in the promoter of 2S protein from *Brassica* species. The RY element plays a key role in tissuespecific and temporal expression of beta-conglycinin gene in coordination with other *cis*-acting elements. Mutant promoter and deletion experiments suggest that RY element involves negative regulation in seeds. A sequence motif CACGTG called 'G-box' was identified as a highly conserved protein-binding site upstream of many genes including a small subunit gene of ribulose*bis*-phosphate carboxylase, *Brassica* 2S protein gene120 and many other genes. Many positive and negative *cis*-regulatory elements were detected in 2S napin promoter of *B. juncea* and *B. napus*. Though these elements are recognised by transacting transcription factors both in transgenic tobacco and rape, some differences in tissue specificity of expression in these two plants were noted.

Expression of Seed Protein Genes

Though conventional breeding techniques will continue to be extensively employed, genetic engineering of crop plants becomes the only option when specific, modified genes are to be transferred between widely related species. The first transgenic plant was reported in 1983. Since then several reviews have been published on the introduction of heterologous/modified seed protein genes of agronomic importance.

There are several methods for transferring genes of interest into plants (Table 2). Of these methods, several techniques like macroinjection into floral

Table 2. Methods of gene transfer to plants

Method	Nature of gene transferred
Agrobacterium tumefaciens	Any gene or DNA fragment between
A. rhizogenes	T-DNA borders
(a) Vector-mediated	
(b) Virus-mediated (Agroinfection)	
Somatic hybridisation	Entire genome by nuclear fusion (hybrid) or cytoplasmic fusion (cybrid)
Direct transfer to protoplast	Segment of DNA or plasmid
(a) Electroporation	
(b) PEG-mediated	
(c) Liposome fusion	
(d) Microinjection	
Biolistic (particle gun)	Segment of DNA or plasmid
Other methods	Segment of DNA or plasmid
(a) Whole seed/seedling uptake	
(b) Vacuum infiltration	
(c) Macroinjection to floral tillers or ovary	
(d) Delivery through stigma/pollen	

T-DNA, Tumour-inducing DNA; PEG, Polyethylene glycol.

stems or developing ovaries and direct uptake through pollen or germinating seed are not very reproducible. Particle bombardment is initially cost intensive. But the advantage is that it needs less trained manpower, i.e. molecular biologists, as technicians can try this method on a number of plants and tissues. On the other hand, *Agrobacterium* Ti (or Ri) vector-mediated gene delivery is more popular due to its being easy, reproducible and less expensive. The only disadvantage is that it is still generally applicable to dicot plants, although there are reports of successful *Agrobacterium* transformation of cereals using acetosyringone.

Plant transformation with numerous genes clearly established that dicot genes including seed protein genes, can be expressed and correctly processed in monocot plants and vice versa. Thus, all the transcription/translation machinery, regulatory sequences, transacting factors and post-translational processing, modifying and targeting enzymes are homologous or similar in all plants. However, due to codon usage preference the level of expression may be different in different host plants and introduced gene combination. A number of seed proteins have been tissue-specifically expressed and correctly processed in transgenic plants like beta-phaseolin in tobacco, beta-conglycinine in *Petunia, Arabidopsis* 2S albumin in tobacco and *Brassica*. In the author's laboratory, *Brassica* 2S protein gene has been expressed in transgenic tobacco seeds, under its own promoter and CaMV 35S promoter. For ease of transformation, the most common plants to which seed protein genes have been transferred are the model plants, tobacco and *Petunia,* for studying the regulatory elements in the promoter, after tagging with a reporter gene like chloramphenicol acetyl transferase (CAT) or beta-glucuronidase (GUS) or firefly luciferase (LUX).

Attempts to Improve the Nutritional Quality of Seed Proteins

There can be two broad approaches for improving the nutritional quality of seed proteins by rDNA techniques:

(1) Gene alteration, in which a seed protein gene already present and expressed in the plant seed is isolated and its coding sequence is altered by addition, substitution or deletion of nucleotides by site-directed mutagenesis, followed by reintroduction of the modified gene under its own promoter into the plant itself; or (2) Gene addition, in which a new capability or quality is conferred to the seed by introduction of new heterologous genes encoding higher levels of desired amino acids, under the seed's own promoter or more efficient heterologous promoter. Again, in the second approach by using a heterologous gene, an ideal animal protein gene like that of a milk protein introduced into a plant is poorly expressed due to difference in codon preference between the plant and animal in question. Therefore, heterologous plant protein genes with desired amino acid composition have been the choice of plant biotechnologists. After the first

transgenic approach to improve Met content of seed proteins by Saalbach *et al.* in 1988, there has been a large flow of literature on this approach. Of several reviews published in recent years, some comprehensive ones are mentioned here. The promising strategy is to introduce genes encoding sulphur-rich and lysine-rich proteins under the control of an efficient promoter into legumes and cereals, respectively. The following illustrates some attempts to improve major crops.

Improvement of Legumes

An ideal candidate for enhancement of methionine content of legume seeds by this strategy is the seed-specific 2S albumin gene of Brazil nut (*Bertholletia excelsa*). Approximately 18% of the codons of this gene code for methionine and 8% for cysteine. This 2S albumin gene under cauliflower mosaic virus (CaMV) 35S promoter was transferred to tobacco and the grain legume *Vicia narbonensis*. The 2S albumin was expressed in all parts of the plant both in tobacco and the legume and was found to be localised in the vacuoles of the leaf mesophyll cells of transgenic tobacco. In another attempt the 2S albumin gene, under a broad bean (*V. faba*) legumin B4 promoter for seed tissue-specific expression was transferred to *V. narbonensis*. This showed that up to 3% of soluble seed protein was 2S albumin. Some transformants had 3-fold increase in methionine content of the salt-soluble pro-tein fraction in seed. To improve the 2S albumin protein of Brazil nut which is sulphur-rich but lacks the essential amino acid tryptophan, residues of tryptophan were inserted at sites without altering the proper protein structure. Three constructs, one with 5 consecutive tryptophan codons and the other two modified genes encoding proteins carrying single tryptophan residues under 35S CaMV promoter were used and all three genes were correctly translated. Several patents claim the enhancement of methionine content of soybean, alfalfa and lotus by the introduction of rice prolamin, sunflower sulphur-rich 2S albumin and maize zein genes. In another attempt, a maize gamma-zein gene encoding a sulphur amino acid-rich protein was used to transform alfalfa and trefoil (*Lotus corniculatus*) under CaMV 35S promoter and RUBISCO small subunit promoter. Expression level was rather low to the extent of 0.05% of alcohol soluble protein. In yet another attempt, gamma-zein gene was modified by the introduction of (pro-lys)*n* coding sequences at different positions. High levels of protein expression were observed when (pro-lys)*n* sequence was contiguous or in substitution of pro-X region of the gamma-zein. Muntz *et al.* summarised the results of genetic engineering for high methionine grain legumes. The foreign protein amounted to 5–10% of the seed proteins in the best transgenic lines. Transgenic soybean reached 100% and narbon bean and lupin reached 80% of the WHO standard for nutritionally balanced proteins. These results document that the Met problem of grain legumes can be solved by genetic engineering.

S-rich 2S albumin gene from sunflower has been introduced into the forage crop clover (*Trifolium subterraneum*) under CaMV 35S promoter along with a sequence encoding ER-retention-signal added to the 3¢-end of the coding region. This prevented the translocation of the expressed protein to the vacuole. Under this condition, the protein was expressed to a maximum level of 0.3% of the total extractable protein in mature leaves.

Improvement of Cereals

Most of the early work on maize and barley improvement was based on the characterisation of 'high lysine' and 'Highproly' lines identified by screening of several thousand collections from wild germplasm. This Highproly mutant has reduced proportion of prolamin (which is deficient in lysine) with an accompanying increase in other more lysine-rich proteins. However, this mutation is highly linked to an agronomically harmful trait, namely, reduced starch content, thereby decreasing grain yield drastically. Attempts to disrupt this linkage by classical breeding did not seem promising. An alternate approach is to introduce a gene for proteins having nutritionally enriched amino-acid composition under strong endosperm-specific promoter like that for high molecular weight glutelin. For example, beta-phaseolin (6% lysine) gene of common bean *Phaseolus vulgaris* under rice glutelin promoter was expressed to the level of 4% of total endosperm protein in transgenic rice seeds.

Mutant or less abundant seed storage protein zein from maize has been frequently used for quality improvement. To increase methionine level, a new methionine-rich zein, normally expressed at low levels was expressed at a high level using the 27 kDa zein promoter. This protein called the high sulphur zein (HS 7) was 21 kDa and contained 37% Met. The level of expression in the transgenic plant was not mentioned. Soybean ferritin gene under rice glu-B-1 promoter has been introduced into rice to increase the iron content. The glutelins together with gliadins are the most abundant proteins in cereal seeds. The high molecular weight glutelin subunit is important in determining the bread-making quality of wheat flour. To change the amount and composition of these proteins, a gene encoding a novel hybrid subunit under the control of native HMW-GS regulatory sequence was inserted into wheat. The hybrid subunit was accumulated to levels comparable to that of native HMW-GS. These studies demonstrate the feasibility of constructing wheat and other cereals with altered levels and types of seed storage proteins.

Improvement of crucifers and other seeds

The small mol. wt. albumin (2S protein or napin) of rapeseeds/ mustard is one of the most important targets for the improvement of its nutritional quality and also for 'Molecular Pharming', i.e. for producing medically important proteins in transgenic plants. There have been several promising attempts for the genetic engineering of 2S albumins. The 5¢-upstream sequences of *Arabidopsis, B. napus,* and *B. juncea* 2S protein confer seed specificity of expression of an introduced gene. These studies demonstrated that 2S protein is correctly expressed and targeted even in a heterologous host. Brazil nut 2S protein gene and a methionine-enriched modified napin were introduced into *B. napus* and were expressed to marginally improve the methionine content of total seed protein. Altenbach *et al.* reported the highest level of expression of Brazil nut protein in transgenic *B. napus* ranging from 1.7% to 4% of total seed protein. This 4% methionine-rich protein contributed about 33% more methionine in the total seed protein. In all these studies, heterologous promoters like that of phaseoline, *Arabidopsis* AT2S1 and soybean lectin were used. The different levels of expression might be due to different promoter efficacy. Studies on *B. juncea* 2S promoter dissection indicated that a suitably modified (to exclude negative regulatory regions and by duplication of enhancer sequence) homologous promoter is more efficient in the expression of a reporter gene. Also, the 3¢-terminator sequences used have always been from a heterologous source. A homologous 3¢-sequence might be better for expression of foreign/modified gene.

In another attempt, an indirect strategy was used in which the antisense gene for cruciferin was introduced into *B. napus,* which resulted in the suppression of cruciferin and increase in the relative proportion of napin, a more balanced protein. The antisense construct contained a part of *cruA* gene in reversed orientation, driven by 5¢-flanking sequence of napin gene, such that the antisense cruciferin mRNA was expressed in a seed-specific manner. The transgenic T1 generation seeds contained 10%, 8% and 32% higher Lys, Met and Cys, respectively, compared to their respective levels in control seeds.

Most seed storage practices have developed empirically, often as a result of a combination of chance observation and pragmatic decisions on what can be achieved most easily by way of environmental control or modification. In agriculture, for example, the driving force for such empirical knowledge has been the simple practical requirement to store seeds reliably from harvest to the next production season. In silviculture, similarly, advice on storing tree seeds has long been available. For example, one and a half millenia ago it was recommended with regard to chestnut (probably *Castanea mollissima*) that "the fresh seeds should be stored in damp soil placed in the house, and during transport the seeds should be packed in a leather bag, because chestnut seeds will die when exposed to sun and wind".

SEED STORAGE BEHAVIOUR

Although far from complete, the world literature includes several comprehensive compendia which provide information on tree seed storage. Some compendia on seed storage behaviour were based on the duration of seed survival (whether air-dried or in soil or water) without considering the role of environmental factors on seed longevity. It was later found that certain species' seeds classified initially as "microbiotic" by Ewart or "short-lived" seeds can in fact be stored for long periods of time under more suitable conditions. Therefore, later compendia have classified species on the basis of seed desiccation tolerance and/or the response of seed survival or longevity to environment.

There are three different, discrete categories having been identified: orthodox seed storage behaviour; recalcitrant seed storage behaviour; and intermediate seed storage behaviour.

Orthodox Seeds

Orthodox seeds can be dried without damage to low moisture contents and, over a wide range of environments, their longevity increases with decreases in seed storage moisture content and temperature in a quantifiable and predictable way. The latter is defined by the seed viability equation

$$v = K_i - p / 10^{K_E} - C_W \log_{10} m - C_H t - C_Q t^2, \qquad (1)$$

where v is probit percentage viability after p days in storage at m percent moisture content (w.b.), t°C, K_i is a constant specific to the seed lot, and K_E, C_W, C_H and C_Q are species viability constants. The constant Ki indicates the initial viability (in probits) of the seed accession when placed in storage. The value of K_i may vary considerably among different seed lots within a species, for example as result of differences in seed production environment, time of harvest, or genotype. The constants C_H and C_Q together describe the response of seed longevity to temperature.

The particular form of the relation between longevity and temperature is a negative curvilinear semi-logarithmic relation whereby the relative benefit to longevity of each 10 °C reduction in temperature declines, the cooler the temperature. A comparison of contrasting species, which included one tree species, found that the values of the temperature terms C_H and C_Q did not differ significantly among species. Similarly, the values of C_H and C_Q determined for four forest tree species (*Liquidambar styraciflua* L., *Pinus elliottii, Pinus taeda,* and *Platanus occidentalis*) by Bonner were 0.0306-0.0508 and 0.000328-0.000976, respectively, and thus close to the common estimates for diverse species of 0.0329 and 0.000478, respectively, determined previously. This suggests that seed longevity in all orthodox species may well show a similar quantitative response to storage temperature.

At one temperature, the gradient of the negative logarithmic relation between seed moisture content (m) and longevity provides the value of the constant C_W. Thus the value of this constant quantifies the effect of moisture content on longevity. The value of C_W differs considerably among species. For example, the values of C_W collated by Hong and others for 53 species from 24 families vary from 0.983 for *Pinus elliottii* to 6.305 for *Sorghum bicolor*. In general, species having seeds with a high starch content, e.g. cereals, have high values for C_W (around 6) but in oily seeds it is much lower, e.g. between 3.5 and 4 for onion (*Allium cepa*) and soyabean (*Glycine max*).

Since tree seeds are generally rich in oil in the embryo and/or endosperm, tree species tend to have lower values for C_W, e.g. 0.983 for *Pinus elliottii* and 4.23 for Norway maple (*Acer platanoides*). In one comparison of estimates of C_W, the mean value for forest tree species was 2.8 compared with a mean value of 4.72 for herbaceous species. One consequence is that to obtain the same relative increase in longevity it is necessary to dry orthodox tree and oily seeds more than starchy seeds from a given initial moisture content.

As a consequence of the negative logarithmic relation between seed moisture content and longevity, the relative benefit to longevity becomes greater for each successive reduction in moisture content. For example, based on the viability parameters estimated for *Ulmus carpinifolia* seeds by Tompsett, the effect of a reduction from 7 percent to 5 percent moisture content provides about a 2.7-fold increase in longevity whereas a reduction from 5 percent to 3 percent in seed moisture content increases seed longevity about 4.5-fold.

There appears, however, to be a common response of longevity to equilibrium relative humidity among contrasting crop species, at least, whereby longevity increases by a factor of about 2.2 for each 10 percent reduction in equilibrium relative humidity. However, we do not yet know whether this generalisation for seeds of arable and horticultural crops also applies to tree seeds.

There are two limits to the negative logarithmic relation between seed moisture content and seed longevity. One is an upper limit, above which seed longevity in hermetic storage is no longer reduced with further increases in moisture, and above which in aerated storage seed longevity increases with further increase in moisture content. The other is a lower limit, below which further reduction in moisture content no longer increases longevity in hermetic storage. The upper moisture-content limit appears to be 11 to 12 percent in pine (*Pinus elliottii* and *Pinus taeda*), 22 percent in elm (*Ulmus carpinifolia*) and *Araucaria columnaris*, and between 15 percent in lettuce (*Lactuca sativa*) to about 26 percent in durum wheat (*Triticum durum*) for crop species. Despite wide variation among species in terms of moisture content, these values coincide with a water potential of about -14

MPa, i.e. the upper moisture-content limit to the viability equation occurs when seed moisture contents are in equilibrium with about 90 percent relative humidity (r.h.) at 20°C.

The lower limit varies substantially between crop species, e.g. between about 6 percent moisture content for pea (*Pisum sativum*) and mungbean (*Vigna radiata*), and about 2 percent for sunflower (*Helianthus annuus*). However, these variant moisture contents coincide with 10 to 12 percent equilibrium relative humidity at 20°C or with a seed water potential of about -350 MPa at this temperature.

In open storage (i.e. where seeds are exposed directly or indirectly to ambient relative humidity; e.g. in a paper bag, cloth bag, or thin, unsealed polythene bag which permits seed moisture content to be influenced by ambient relative humidity) at ambient temperature the longevity of orthodox seeds varies considerably. For example, viability of *Salix* spp. cannot be maintained beyond 10 days in open storage while seeds of *Cassia multijuga* were capable of 100 percent germination when tested after 158 years in open storage at room temperature in a Mediterranean climate.

For the maintenance of viability in long-term storage, *Salix* spp. require hermetic storage at low moisture contents and cool temperatures, e.g. sub-zero temperatures and moisture contents in equilibrium with 10 to 30 percent r.h. (relative humidity). Clearly, seeds of *Cassia multijuga* do not need such precise control of the storage environment in order to maintain viability in storage over the long term.

Provided estimates of the viability constants are available, the viability equation can be applied to determine suitable air-dry storage environments for a given duration of storage for a particular orthodox species. Since the same storage life can be achieved by different combinations of storage temperature and seed moisture content–the precise values depending upon species–there can be considerable flexibility.

Since the relative benefit to longevity from reducing temperature becomes less the lower the storage temperature, while the relative benefit to longevity from reducing moisture content becomes greater the lower the moisture content (within the wide range of environments over which the viability equation applies), it is often more cost-effective to reduce seed moisture content than to lower storage temperature.

This approach is of particular importance for seed storage centers where refrigeration to zero and sub-zero temperatures cannot be provided. In such situations it has been recommended that the seeds be dried to moisture contents in equilibrium with 10 to 12 percent r.h. at 20 °C and then stored hermetically at ambient or (preferably) cooler temperatures. This low-technology approach has been described by some as "ultra-dry" seed storage. Results from a 5-year investigation have shown that the longevity of ultra-dry seed at ambient temperatures is greater than that of crop seeds stored at 5 to 6 percent moisture content, at least for several crops. Similarly, over a much longer duration, 90 percent germination of seeds of barley (*Hordeum vulgare*) was reported when tested after 110 years in ultra-dry storage (3.1 percent moisture content) at 10 to 15°C. However, to date, ultra-dry seed storage has not been applied to tree seeds to our knowledge.

Successful storage of orthodox tree seeds over the long term under the FAO/IPGRI preferred conditions, i.e. 3 to 7 percent moisture content and -18 °C, has been realised for more than two decades in the Royal Botanic Gardens' Kew Seedbank and for about 7 years at the International Institute of Tropical Agriculture. For example, seeds of *Rhus verniciflua* and *Acacia tortilis* stored in the Kew Seedbank since 1970 and 1974, respectively, showed 100 percent germination in recent tests.

Long-term storage of orthodox tree seeds has also been achieved by cryopreservation (or cryostorage) at ultra-low temperatures from -80 °C to -196 °C with liquid nitrogen. For example, no loss in viability was reported for *Pinus ponderosa, Pseudotsuga menziesii, Thuja plicata,* and *Tsuga heterophylla* following 3 years of cryopreservation in liquid nitrogen. An essential first step in seed cryopreservation is the determination of optimum (safe) moisture contents for each orthodox species, particularly those with oily seeds. Wang and others collated information on the cryopreservation of orthodox tree seeds; it suggests that moisture contents between 3.8 and 11 percent are safe for at least short periods of time (from 4 days to 3 years).

However, a loss in viability of about 10 percent following 14 days storage in liquid nitrogen at 6 to 8 percent moisture content was reported for seeds of certain tree species. One possible cause of problems with liquid nitrogen seed storage is the requirement to optimise both the rate of cooling and the subsequent rate of warming; these must be determined empirically. Pretreatment with cryoprotectants sometimes improves seed survival in liquid nitrogen, but in some circumstances can be damaging. Mechanical injury to seed by direct immersion into liquid nitrogen causing loss in viability has also been reported for some species. Finally, despite suggestions of infinite seed longevity by cryopreservation in liquid nitrogen, there is no firm evidence that reducing seed storage temperatures much below -20 °C results in further significant benefits to longevity, at least for tree seeds. For example, Tompsett found similar longevity for *Ulmus carpinifolia* seed stored at -13 °C and -75 °C.

For the maintenance of viability over the medium term, i.e. a period of 2 to 10 years or so, hermetic storage in a wide range of contrasting environments can be successful. For example, with relatively high moisture contents (in the air-dry range) combined with cool temperatures, e.g. 8 to 10 percent moisture content and 0 to 10 °C for most leguminous trees, or low moisture contents combined with relatively warm temperatures, e.g. 3 to 7 percent moisture content and 10 °C to ambient temperatures, e.g. Barton reported 33 percent germination for *Ulmus americana* L. seeds following 4 years of hermetic storage at 3 percent moisture content at room temperature. Ultra-dry seed storage may be applicable to tree seeds for medium-term storage at ambient temperatures. For example, Joseph and Brinkman recommended that seeds of *Betula* spp. be stored at moisture contents as low as 0.6 to 1 percent at room temperature.

In addition to air-dry storage over a very wide range of different environments, successful storage of orthodox seeds over the short-term, from a few months to 1 or 2 years, can also be achieved by moist storage at cool temperatures. This approach is commonly used for tree species that exhibit strong dormancy such as those adapted to temperate climates (i.e. temperate latitudes and high altitudes of the Tropics) since prechilling during storage can remove dormancy. For example, several orthodox tree species of temperate climates, such as *Carya* spp., *Chionanthus virginicus, Euonymus* spp., *Hamamelis virginiana, Ilex* spp., *Juglans* spp., *Juniperus* spp., *Liriodendron tulipifera, Nyssa* spp., *Picea* spp., and *Taxus* spp. can be stored moist at 3 to 5 °C.

Similarly, orthodox seeds of several species that are native to high altitudes in the Tropics and show considerable dormancy can also be stored moist at 3 to 5 °C. For example, seeds of *Prunus africana,* native to high mountains between 900 m and 3400 m in Africa, and *Michelia champaca* L. and *Michelia compressa,* native to high mountains of Asia, can be stored moist at 4 °C for 1 year. In addition, seeds of some pioneer forest trees that have an absolute requirement for light in order to germinate can also be stored moist in the absence of light at ambient temperature for long periods of time. For example, seeds of *Piper aequale* can be maintained for 7 years fully imbibed in darkness without loss in viability.

Recalcitrant Seeds

Recalcitrant seeds cannot be dried without damage and so their longevity does not respond to air-dry environment in the manner described by the seed viability equation (1). When freshly harvested recalcitrant seeds are dried, viability is at first slightly reduced as moisture is lost, but then begins to be reduced considerably at a certain moisture content, termed the "critical moisture content" or "lowest safe moisture content". If drying continues further, viability is eventually reduced to zero. Critical moisture contents for loss of viability on desiccation vary greatly among the recalcitrant species, among cultivars and seed lots, and depending upon the stage of seed maturity at time of collection.

The critical moisture content may also vary with the method of seed drying. The values of the "lowest safe moisture content" vary between extremes of about 23 percent for cocoa (*Theobroma cacao*) to 61.5 percent for

Avicennia marina. Despite this variation, these moisture contents are equivalent to a relatively narrow band of relative humidities of 96 to 98 percent, or seed water potentials of about -1.5 MPa to -5 MPa.

There is no satisfactory method of maintaining the viability of recalcitrant seeds over the medium and long term. Since they cannot be dried neither can they be stored at subzero temperatures, because they would be then killed by freezing injury resulting from ice formation. The longevity of recalcitrant seeds is short, from a few weeks to a few months for species adapted to tropical environments, and up to about 3 years for several species adapted to temperate environments. However, if the optimum storage environments are carefully determined, longevity of several tropical recalcitrant seeds can be extended to 3 years, at least for species of *Symphonia globulifera*.

The principle of successful moist seed storage for recalcitrant seeds is that seeds must be maintained at moisture contents close to that at which they are shed, with continuous access to oxygen; these circumstances minimise seed deterioration since repair mechanisms can operate. Under these conditions (high seed moisture and available oxygen), however, seeds tend to germinate. Clearly it is also essential that the conditions should prevent or at least delay germination. It is therefore easier to store recalcitrant species with dormant seeds (either primary dormancy or induced dormancy) than with nondormant seeds under such conditions.

For nondormant seeds, as shown by most tropical tree seeds at maturity, low temperatures can reduce the rates of both seed deterioration and germination provided they remain above the value that results in chilling damage or the lower value at which ice crystallisation occurs. For example, chilling injury occurs at between 5 and 10 °C for *Shorea roxburghii*, 5 and 12 °C for *Symphonia globulifera* and *Hopea odorata* Roxb. Determination of the optimum temperature for imbibed seed storage is required for each recalcitrant species. A suggested protocol for determining these values has been described elsewhere.

The optimum temperature appears to be that at which nondormant seeds remain alive but are unable to germinate, i.e. the base temperature for germination. For example, the optimum storage temperatures determined in this way are 10 °C for *Shorea roxburghii*, 12 °C for *Mangifera indica*, and 15 °C for *Hopea odorata* and *Symphonia globulifera*. However, recalcitrant seeds of species adapted to temperate climates show considerable dormancy which requires long periods of prechilling at 2 to 5°C to overcome. Furthermore, seeds of many temperate species with recalcitrant seed storage behaviour have the ability to germinate at 2 to 5 °C, and thus moist storage of such recalcitrant seeds at 0 to 10 °C may result in germination during storage.

In such cases, germination may be prevented by reducing seed moisture content slightly (by about 5 percent below that of fresh seeds), or either reducing the storage temperature to below the optimum prechilling temperature (e.g. 0 to -3 °C) or increasing it to the base temperature for germination of dormant seeds. For example, the viability of seeds of sycamore (*Acer pseudoplatanus*) and *Quercus robur* can be maintained for 3 years at 24 to 32 percent and 40 to 45 percent moisture content, respectively, at -1 to -3 °C, and horsechestnut (*Aesculus hippocastanum*) for 3 years if moist dormant seeds are stored at 16 °C, although in this case subsequent prechilling is required to remove dormancy.

In practical terms, species with recalcitrant seeds can therefore be divided into two subcategories: those adapted to tropical climates, and those adapted to temperate climates (i.e. temperate latitudes, or high altitudes in the Tropics). In general, the viability of recalcitrant seeds can be maintained (albeit only for limited periods in aerated conditions) at moisture contents just less than fully imbibed, i.e. about 2 to 5 percent below those of fresh seeds, or in equilibrium with 98 to 99 percent r.h., at optimum storage temperatures which vary from about 7 °C to 17 °C among species adapted to tropical climates, and between about -3 °C and 5 °C among many of those adapted to temperate climates.

Maintaining recalcitrant seeds fully imbibed or subimbibed, with continuous aeration, and at the same time preventing germination and fungal contamination, is difficult. Aeration can result in loss of seed moisture, and respiration can quickly deplete oxygen. The storage medium is therefore very important for recalcitrant seeds. It should fulfill two functions: first, to maintain seed moisture constant at high values; second, to allow diffusion of sufficient oxygen to the moist seeds. The storage of moist recalcitrant seeds in damp charcoal, sawdust, or moist sand is generally reported to be more efficient than storage in polyethlene bags.

Storage at or near the harvest moisture con-tent in media such as sawdust (at 16 percent moisture content) or perlite (at 0 to 4 percent moisture content) within suitable containers, such as open-weave sacks or bags, placed in a highhumidity room has been recommended. Coating recalcitrant seeds with alginate or alginate with ABA was also reported to extend seed lifespan by a factor of up to 3 or 4.

Cryopreservation of zygotic embryos has been reported to be successful for a number of tree species that show orthodox, intermediate, or recalcitrant seed storage behaviour. For successful cryopreservation, excised embryos or embryonic axes must survive desiccation below the threshold freezable moisture content of about 18 to 33 percent, below which value there is no freezable water for ice formation in ultra-low temperatures. Embryos and embryonic axes of some recalcitrant species are able to survive desiccation to lower moisture contents than are the whole seeds.

For example, fresh seeds (36 percent moisture content) of *Hevea brasiliensis* tolerated desiccation to 20 percent moisture content, but no seeds survived further desiccation to 15 percent moisture content. However, after desiccation to 16 percent moisture content followed by 16 hours cryopreservation in liquid nitrogen, 87 percent and 69 percent, respectively, of the excised embryonic axes (55 percent moisture content) survived. Unfortunately, embryos of many recalcitrant seeds are damaged by desiccation before seed moisture content is reduced below those values at which ice crystallisation no longer occurs, and so the cryopreservation of excised embryos of these species results in death, either by ice formation for embryos at moisture contents above the threshold freezable moisture content, or by desiccation to moisture contents below that threshold.

Fast drying (also termed "flash" drying) may allow excised embryos to survive desiccation to lower moisture contents than those dried more slowly within intact seeds. Excised embryos or embryonic axes can be dried quickly in a laminar airflow cabinet at room temperature or under a flow of compressed air at about 9 liters min^{-1}. After flash drying of embryonic axes extracted from mature seeds of *Landolphia kirkii* reduced moisture content from 67 percent to 23 percent in 30 minutes, 90 percent of the excised embryonic axes were able to germinate. Fast drying with silica gel or with an aseptic air current has allowed excised embryonic axes to survive desiccation to a lower value than that achieved by vacuum drying.

The stage of maturity of the seeds from which the embryos are excised is an important factor influencing survival following desiccation: embryos excised from mature seeds not only survived desiccation to lower moisture contents than immature embryos, but also survived subsequent cryopreservation in liquid nitrogen. Fast freezing can allow desiccated embryos to better survive cryopreservation in liquid nitrogen. In contrast, slow freezing with immature excised embryos of cocoa has been reported to be more successful than fast freezing. The specimen to be frozen must be as small as possible, and pretreatment of embryos with cryoprotectants, such as DMSO, proline, sucrose, and glycerol, are important factors influencing the survival of excised embryos (or embryonic axes) in liquid nitrogen.

Results reported for *Aesculus* spp., *Araucaria hunsteinii, Artocarpus heterophyllus* Lam., *Castanea sativa, Citrus hystrix, Coffea liberica, Cocos nucifera, Dimocarpus longan, Hevea brasiliensis, Landolphia kirkii, Quercus* spp., *Trichilia dregeana* and *Theobroma cacao* have shown the feasibility of the cryopreservation of excised embryos or embryonic axes of recalcitrant seeds. However, others have reported a lack of success using the same species. For example, experiments with excised embryos of *Artocarpus heterophyllus, Landolphia kirkii,* and *Quercus robur* found that none survived cryopreservation in liquid nitrogen.

Intermediate Seeds

Although the term "intermediate seed storage behaviour" was introduced only at the beginning of this decade, this third category of seed storage behaviour was hinted at more than eight decades ago. Elliott divided temperate forest tree seeds into three classes:

i) those that can be dried,

ii) those that can survive partial drying, and

iii) those that can rarely be dried at all.

The definition of intermediate seed storage behaviour is based on the response of longevity to storage environment. In seeds showing intermediate seed storage behaviour, the trend toward increasing longevity the lower the seed storage moisture content (within the air-dry range) is reversed at a comparatively high moisture content, whereby at lower moisture contents longevity is reduced. But in such species it is also often (but not always) observed that seeds may be damaged immediately by desiccation to relatively low moisture contents, about 7 to 12 percent moisture content depending on species.

The critical moisture contents of intermediate seeds below which more rapid loss in viability occurs during hermetic storage varies considerably with species, degree of maturity, and method of seed extraction and/or postcollection handling. In general, seeds extracted from fruits at maturity tolerate desiccation to moisture contents in equilibrium with about 40 to 50 percent r.h., i.e., about 10 percent moisture content for arabica coffee (*Coffea arabica*) and 7 percent moisture content for *Citrus* spp.

A further feature of intermediate seeds of tropical origin is the fact that the longevity of dry seeds (7 to 10 percent moisture content) is reduced with reduction in storage temperature below about 10 °C. In such cases, then, there is an optimum air-dry storage environment for the maintenance of seed viability. In arabica coffee this is about 10 °C with 10 to 11 percent moisture content. It is now known that seeds of neem (*Azadirachta indica* A. Juss), *Cinnamomum subavenium*, several *Citrus* species, *Coffea arabica*, robusta coffee (*Coffea canephora*), oil palm (*Elaeis guineensis*), Khaya senegalensis, Lindera megaphylla, Neolitsea parvigemma, and Swietenia *macrophylla* show intermediate seed storage behaviour.

Among about 7,000 species examined in a recent survey, about 134 species are believed to show intermediate seed storage behaviour. Many of these species are trees of tropical origin, but a few are herbaceous. For example, that survey suggests that the following important tree species may show intermediate seed storage behaviour: Acer macrophyllum, Afrocarpus gracilior, Agathis macrophylla, Araucaria columnaris, A. nemorosa, A. rulei, A. scopulorum, Bertholettia excelsa, Bixa orellana, Chrysophyllum cainito,

Dacrycarpus dacrydioides, Dipterocarpus alatus Roxb. & G. Don, D. intricatus, D. tuberculatus, Dipteryx odorata, Rhapidophyllum hystrix, and Virola surinamensis.

From the point of view of optimum air-dry seed storage environments, it may be helpful to distinguish between species with intermediate seed storage behaviour adapted to tropical environments and those adapted to temperate environments. For example, intermediate seeds of tropical origin such as arabica coffee and papaya (*Carica papaya*) can be stored at moisture contents in equilibrium with 50 percent r.h. (9 to 10 percent moisture content) and 10 °C for up to 5 and 6 years, respectively, without loss in viability.

The viability of intermediate seeds of temperate origin is also maintained well at moisture contents in equilibrium with about 50 percent r.h., but at cooler temperatures of 5 °C to -10 °C. For example, seeds of *Corylus avellana* can be stored hermetically at -5 °C with 10 to 13 percent moisture content. Species showing intermediate seed storage behaviour can be stored successfully over the medium term, therefore, provided optimum environments have been defined and can be maintained.

Like orthodox seeds, intermediate seeds can also be stored moist at cool temperatures if germination is either prevented or delayed. For example, seeds of arabica coffee and tea (*Camellia sinensis*) were stored moist at 15 °C and 1 °C, respectively, for 2.5 and 6 years, respectively, with little loss in viability. The situation regarding the cryopreservation of intermediate seeds is a little confused. Despite reports of the immediate death of (whole) seeds of species with intermediate seed storage behaviour following cryopreservation in liquid nitrogen, e.g. arabica coffee, *Corylus avellana, Corylus cornuta,* oil palm, and *Roystonea regia* (Kunth) O.F. Cook, there have been several reports of the survival of dry seeds of species with intermediate seed storage behaviour following immersion in liquid nitrogen, e.g. *Camellia sinensis, Carica papaya, Citrus aurantifolia,* and *Citrus halimii*. This implies that it may be possible to improve cryopreservation techniques to make them suitable for intermediate seeds.

Since the whole seeds of species with intermediate seed storage behaviour tolerate desiccation to relatively low moisture content (7 to 10 percent), desiccated embryos may have a greater chance to survive cryopreservation in liquid nitrogen than is the case for recalcitrant seeds. Reports of successful cryostorage of excised embryos of *Azadirachta indica, Camellia sinensis, Citrus aurantifolia* and *Citrus halimii, Coffea arabica, Corylus avellana, Elaeis guineensis, Howea forsteriana, Poncirus trifoliata,* and *Veitchia merrilli* have shown the potential for long-term seed storage under such conditions.

DESICCATION TOLERANCE AND POTENTIAL SEED LONGEVITY

Application of the seed viability equation (1) reveals that initial seed quality

(K_i)–potential longevity–plays an important role in ensuring good seed survival in storage. The higher the value of K_i the longer the period of viability for a given storage environment. During seed development and maturation, the value of the seed lot constant K_I–so the potential longevity of orthodox seeds–increases markedly.

There is now good evidence in a wide range of contrasting orthodox crop species that potential longevity continues to increase during the period after the end of seed filling, as seeds mature. There is also good evidence that the ability of orthodox seeds to tolerate desiccation to very low moisture contents also improves considerably during the late stages of seed maturation, and indeed one study has, therefore, shown a strong association in developing and maturing seeds between potential longevity (K_i) and the ability of seeds to tolerate desiccation to very low moisture contents.

The ability of seeds to tolerate enforced rapid desiccation has been reported to increase during seed development and maturation of tree seeds with orthodox, intermediate, and recalcitrant seed storage behaviour. Orthodox seeds of several crop, wild herb, and tree species are not able to tolerate rapid enforced desiccation to low moisture contents (4 to 5 percent) until some time after mass maturity once maturation drying has reduced seed moisture content on the mother plant substantially.

Maximum tolerance of rapid enforced desiccation occurs some time after mass maturity, probably at natural dehiscence. In addition, the slow desiccation of immature seeds harvested before mass maturity, and well before maturation drying occurs naturally, enables tolerance of rapid desiccation to be acquired. For example, in Norway maple (*Acer platanoides*), rapid drying (by reducing moisture content from 57.3 percent to 9 percent within 1 day, and to 5 percent within 4 days) of the seeds harvested at mass maturity reduced viability from 100 percent to 61 percent and 38 percent at 8.1 percent and 4.7 percent moisture content, respectively, while for seeds that were first dried slowly for 32 days, during which period seed moisture content was reduced from 57.3 percent to 29.9 percent, subsequent rapid drying to 3.5 percent moisture content (within 4 days) resulted in 93 percent viability. The slow drying of seeds or fruits harvested before or at mass maturity has also been reported to increase seed vigor and potential longevity in several crop and wild species.

In addition, desiccation tolerance also decreases when seeds are initially subjected to conditions under which germination is initiated. For example, prechilling, moist storage, presoaking, fermentation treatment for seed extraction, and seed priming can reduce desiccation tolerance, and hence alter seed storage behaviour. Similarly, seeds that are produced in unfavourable environments may show reduced desiccation tolerance to low moisture contents. For example, seeds of a japonica rice (*Oryza sativa* subsp. *japonica*) produced in a hot environment of 32/24°C (day/night) showed consistently poorer desiccation tolerance at each stage of maturation drying than those produced in a cooler environment of 28/20 °C.

Prediction of Seeds

Methods, and indeed feasibility, of seed storage depend upon the seed storage behaviour shown by a particular species. A simple, two-stage protocol to determine seed storage behaviour has been suggested elsewhere. This task is, however, considerable given that there are about 250,000 species of flowering plants. Estimation of likely seed storage behaviour would therefore be helpful, and several approaches to doing this have been developed. No single criterion can provide a satisfactory estimation of likely seed storage behaviour, but combining information for at least four of the six factors described below may be worthwhile.

PLANT ECOLOGY AND SEED STORAGE BEHAVIOUR

There appears to be an association between plant ecology and seed storage behaviour. From information on seed storage behaviour collated for almost 7,000 species from 251 families, it is evident that species that show recalcitrant seed storage behaviour do not occur naturally in (i.e. originate from) arid habitats, that is desert and savanna. In such environments, the majority of plant species show orthodox seed storage behaviour, while a few may show intermediate seed storage behaviour. It is clear, however, that further generalisations are not possible. In particular, for example, at the other ecological extreme it is quite clear that not all species native to moist habitats, rain forests, flooded forests, or aquatic environments show recalcitrant seed storage behaviour; all three categories of seed storage behaviour can be found among the species native to such moist environments.

Taxonomic Classification

Orthodox seed storage behaviour can be found in all species within a few families (e.g. Chenopodiaceae, Combretaceae, Compositae, Labiatae, Solanaceae, Pinaceae), or almost all species in others with only a few exceptional species (e.g. Leguminosae, Graminae, Cucurbitaceae, Cruciferae and Rosaceae). However, most families of flowering plants including tree species have species with two (e.g. Anacardiaceae, Dipterocarpaceae, Euphorbiaceae, etc.) or three different categories of seed storage behaviour (e.g. Meliaceae). Furthermore, seed storage behaviour can differ among species within a genus.

Plant, Fruit, or Seed Characteristcs

Certain generalisations appear possible concerning associations between fruit and seed characteristics and seed storage behaviour. For example, orthodox seed storage behaviour is shown by species that produce achenes, many-seeded berries, many-seeded dehiscent capsules, many dry-seeded pods (but not arillate), many dry-seeded follicles, schizocarps, and utricles. Most species that produce siliques (one exception) and caryopses (three known exceptions) also produce orthodox seeds. On the other hand, however, all three categories of seed storage behaviour can be found among species that produce one seed or a few large seeds (from 1 to 10 seeds) per fruit, or many arillate seeds per fruit, among drupes, pods, capsules, berries, and nuts.

Seed size

Recalcitrant seeds often tend to be larger than intermediate seeds, which in turn tend (on average) to be larger than orthodox seeds. However, there is a very wide range of seed weights or sizes over which seeds of different species show all three types of seed storage behaviour.

Seed moisture content at shedding

A recent survey shows that seed moisture contents at maturity or shedding for species with recalcitrant, intermediate, and orthodox seed storage behaviour overlap considerably. No generalisation on seed storage behaviour is possible for species in which seeds at maturity or shedding have between about 25 percent and 55 percent moisture content.

Seed shape

There appears to be a weak association between seed shape and seed storage behaviour. For example, recalcitrant seeds are often spherical or oval. Seeds that are thin and flat, shapes that aid natural maturation drying, tend to show orthodox seed storage behaviour. Flat seeds tend to show orthodox seed storage behaviour in *Acer* spp. , but show both orthodox and intermediate seed storage behaviour in Meliaceae. The spherical form is present in all three categories of seed storage behaviour.

All of the single criterion suggested above are not sufficiently reliable to be very useful as indicators of likely seed storage behaviour. However, it may be possible to develop a multiple-criteria predictive framework in order to suggest likely seed storage behaviour. For example, combining information on the four criteria–seed weight, seed moisture content at shedding, seed shape, and taxonomy can form the basis of a guide to likely seed storage behaviour among *Acer* spp. Similarly, the four criteria of seed weight, seed moisture content at maturity, seed shape, and plant ecology may provide a guide to the estimation of seed storage behaviour in Meliaceae (fig. 1).

For example, from the two positions designated A in figure 1 for seed weight (TSW = 4300 to 7700 g), ecology (moist evergreen forest of India), seed shape (convex, fleshy aril) of *Amoora wallichii*–information from Luna –we suggest that this species is likely to show recalcitrant seed storage behaviour. Similarly, from the two positions designated B for seed weight, seed shape (convex, fleshy aril), and ecology (rain forest of East Africa) of *Turraeanthus africanus*, we suggest that recalcitrant seed storage behaviour is likely in this species also.

Furthermore, the information available on natural habitat of the two species can also be used to suggest likely suitable moist seed storage conditions. Given that the *Amoora wallichii* tree is described as "frost-hardy", and that *Turraeanthus africanus* grows at an altitude of 1525 m, we suggest that chilling injury is unlikely if the seeds of both species are stored moist at about 10°C.

REFENCRES

International Workshop on Seeds, and G. Nicolas. 2021. *The biology of seeds recent research advances : proceedings of the Seventh International Workshop on Seeds, Salamanca, Spain 2002*. Wallingford, Oxon, UK: CABI Pub. Page 113.

Bewley, J. Derek, and Michael Black. 2014. *Seeds physiology of development and germination. The language of science*. New York: Plenum Press. page 230.

13

Seed Treatment

Seed treatment refers to the application of fungicide, insecticide, or a combination of both, to seeds so as to disinfect and disinfect them from seed-borne or soil-borne pathogenic organisms and storage insects. It also refers to the subjecting of seeds to solar energy exposure, immersion in conditioned water, etc. The seed treatment is done to achieve the following benefits.

Seed treatment:

1) Prevents spread of plant diseases
2) Protects seed from seed rot and seedling blights
3) Improves germination
4) Provides protection from storage insects
5) Controls soil insects.

Presently, 70% requirement of seed is met from the farmer's own stock which go for sowing without seed treatment. Even if seed is sourced from the private or public sector agencies, except hybrid seeds, large percentage of such seed is untreated. The estimates reveal that on an average, 80% of the seed sown in the country is untreated, as against the 100% seed treatment practice in developed countries. Seed treatment not only protects the seeds from seed and soil borne diseases but also gives protection to the emerged seedlings from sucking insect pests affecting crop emergence and its early growth. However, many farmers in the country are neither familiar with the practice nor follow it. The practice of seed treatment is also IPM compatible.

However, seed treatments should not be considered as a cure for all maladies for the selection of poor/unhealthy seed lots. For example, treatment of seeds with excessive mechanical damage or other damages or seeds kept in poor storage conditions, or genetic differences in a variety will not increase seed germination.

Keeping in view of the importance of seed treatment to achieve better crop stand of major crops, and the fact that many farmers in the country do not follow/adopt this practice, adoption of seed treatment by the farmers across the country require effective extension strategies and making the appropriate chemical/bio- products available to the farmers at their doorstep. The farmers also required to be trained/made aware of the methods of seed treatment, post treatment- handling of the seeds and planting materials. Government of India has therefore launched a country wide campaign for ensuring 100% seed treatment in all important crops during coming Rabi season. Pesticide industry associations, ATMAs, CIPMCs, KVKs, Farmers Clubs, SAUs, NGOs, etc.can play an important role in the campaign for 100% seed treatment and hence their participation is solicited.

TYPES OF SEED TREATMENT

1) *Seed disinfection*: Seed disinfection refers to the eradication of fungal spores that have become established within the seed coat, or i more deep-seated tissues. For effective control, the fungicidal treatment must actually penetrate the seed in order to kill the fungus that is present.
2) *Seed disinfestation*: Seed disinfestation refers to the destruction of surface-borne organisms that have contaminated the seed surface but not infected the seed surface. Chemical dips, soaks, fungicides applied as dust, slurry or liquid have been found successful.
3) *Seed Protection*: The purpose of seed protection is to protect the seed and young seedling from organisms in the soil which might otherwise cause decay of the seed before germination.

CONDITIONS UNDER WHICH SEED MUST BE TREATED

1) Injured Seeds: Any break in the seed coat of a seed affords an excellent opportunity for fungi to enter the seed and either kill it, or awaken the seedling that will be produced from it. Seeds suffer mechanical injury during combining and threshing operations, or from being dropped from excessive heights. They may also be injured by weather or improper storage.
2) Diseased seed: Seed may be infected by disease organisms even at the time of harvest, or may become infected during processing, if processed on contaminated machinery or if stored in contaminated containers or warehouses.
3) Undesirable soil conditions: Seeds are sometimes planted under unfavourable soil conditions such as cold and damp soils, or extremely dry soils. Such unfavourable soil conditions may be favourable to the

growth and development of certain fungi spores enabling them to attack and damage the seeds.

4) Disease-free seed: Seeds are invariably infected, by disease organisms ranging from no economic consequence to severe economic consequences. Seed treatment provides a good insurance against diseases, soil-borne organisms and thus affords protection to weak seeds enabling them to germinate and produce seedlings.

PRECAUTIONS IN SEED TREATMENT

Most products used in the treatment of seeds are harmful to humans, but they can also be harmful to seeds. Extreme care is required to ensure that treated seed is never used as human or animal food. To minimise this possibility, treated seed should be clearly labelled as being dangerous, if consumed. The temptation to use unsold treated seed for human or animal feed can be avoided if care is taken to treat only the quantity for which sales are assured.

Care must also be taken to treat seed at the correct dosage rate; applying too much or too little material can be as damaging as never treating at all. Seed with a very high moisture content is very susceptible to injury when treated with some of the concentrated liquid products.

If the seeds are to be treated with bacterial cultures also, the order in which seed treatments should be done shall be as follows:

i) fungicide

ii) bacterial cultures.

Seed Treatments have played and are still playing a significant role in the history of mankind and the ability to stave off hunger and starvation in improving the establishment of healthy crops leading to better yields. Seed Treatments are the use and application of biological, physical and chemical agents and techniques used with seed that provide seed and plant protection and improve the establishment of healthy crops.

Some of the first recorded uses of Seed Treatment are from the Egyptian and Roman periods and involved a technique of using sap from onion (*Allium spp.*). In the middle ages, Seed Treatments were made with liquid manure and chlorine salts. Salt water treatments have been used since the mid-1600s and the first copper products were introduced in the mid-1700s. The "still used today" technology of hot water treatment is documented from as early as 1765 in Wittenberg, Germany. Seed was placed in 45°C water for 2 hours, this provided control of certain fungal pathogens on the outer surface of the seed.

The seed industry is continuously in a time of transition and improvement. Key milestones in the history of Modern Seed Treatment are the introduction and ban of arsenic (used from 1740 until 1808), the introduction and ban of Mercury (used from 1915 until 1982).

The launch of the first systemic compound in 1960; until then, Seed Treatments had been only seed sterilants and did not move into the plant. During the 1970's the first systemic Seed Treatment fungicides for air-borne pathogens were introduced. In the 1990's, new modern classes of fungicide and insecticide chemistry were launched.

The Seed Industry has a long history of not only treating seeds but also of waste management of treated seeds. As early as 1786, there is documentation of a ban against feeding and grinding treated seed, understandable, because arsenic was a popular Seed Treatment in this era.

Today, the waste management of non-usable treated seeds, empty containers, and waste water are priorities for Seed Treaters and Seed Producers. The Seed and Seed Treatment Industries have a long history of partnership and dedication in bringing to the farmer high quality seed. The seed is not only as pest and disease free as possible, but treated to provide the possibility of pest and disease control, when needed, during germination and emergence of the young plant and early growth of the crop.

The objectives of modern Seed Treatment products are the superior control of certain insects and diseases while improving crop safety leading to good establishment of healthy and vigorous plants. Modern Seed Treatment formulations must also contribute to improvements in farmers' and workers' safety and stewardship of the environment.

Today's modern Seed Treatment products have to meet high safety and efficacy standards. The newest active substances and formulations provide long lasting, broad spectrum, systemic control of diseases and insects (depending on the specific active ingredient). Modern formulated Seed Treatment products used by the Farmer, Seed Treater, and Seed Producer are often sophisticated with several active ingredients and special wetting agents and colorants that are safe to the seed, the environment and the user.

Modern Seed Treatment products are able to deliver high levels of efficacy for the control of early season insects or diseases at a much reduced usage rate compared to many foliar or soil applied alternatives. Seed Treatment usage with today's compounds reduces the land surface exposed to the active ingredient. Imagine the comparison of 1 hectare of land with a foliar or soil applied application of active ingredient—results in 10,000 m^2 of land in contact with the active ingredient. If the application was used in-furrow, then the exposed land surface could be reduced to 500m^2. But, the use of Seed Treatment would leave only 58m2 of surface exposed to the active ingredient, minimising environmental impact. Additionally, seed treatments have less risk of impact on non-target organisms and drift.

INTEGRATED PEST MANAGEMENT

The use of Seed Treatment is consistent with the concept of Integrated Pest Management. The farmer must strive for the best pest management method that is safe, environmentally sound, and scientifically proven. Without the use of Seed Treatment, farmers would have great difficulty to control certain seed-borne and early season seedling diseases and insects and would have to resort to more expensive and less environmentally acceptable methods.

The FAO International Code of Conduct on the Distribution and Use of Pesticides (Article 2), states, "Integrated Pest Management (IPM) means a pest management system that, in the context of the associated environment and the population dynamics of the pest species, utilises all suitable techniques and methods in as compatible a manner as possible and maintains the pest populations at levels below those causing economically unacceptable damage or loss." Seed Treatment can be used as a sound basis for a successful Integrated Pest Management Program.

APPLICATION OF SEED TREATMENT

Seed Treatment is a term that describes both products and processes. The usage of specific products and specific techniques can improve the growth environment for the seed, seedling, and young plant. Seed Treatment complexity ranges from a basic dressing, to coatings and pelleting:

- *Seed Dressing*: The most common method of Seed Treatment. The seed is dressed with either a dry formulation or wet treated with a slurry or liquid formulation. Dressings are applied both on-farm and industrially
- *Seed Coating*: A special binder is used with a formulation to enhance adherence to the seed. Coatings require advanced treatment technology.
- *Seed Pelleting*: The most sophisticated Seed Treatment technology, resulting in changing the physical shape of a seed to enhance plantability and handling. Pelleting requires specialised application machinery and techniques and is the most expensive of the applications.

Depending on the type of desired treatment, various treating machines can be used. The more complex the treatment, the more cost is involved. Many modern treatment machines can provide specific formulation and dosing by way of a direct injection process. The most sophisticated treatment machines can provide the various formulations and additives as layers on the seed, which allows precise placement of the active ingredient.

Farmers, distributors, officials, and regulatory authorities have high standards for the Seed Industry regarding the quality of a Seed Treatment and its application to the seed. The expectations are that Seed Treatment will provide high performance and safety in processes and products.

The Seed Industry recognises that Seed Treatment supports "Added Seed Value" (e.g., safe emergence, seedling establishment, crop stands, yield, and crop quality). Seed Treatments complement and protect genetic improvements, function as part of IPM (Integrated Pest Management), are safe and economical in use, and reduce personal and environmental hazards. Seed Treatments, compared to conventional crop protection products, offer competitive costs, reduced application efforts and save time. Seed Treatment is sometimes the only timing and delivery route for the control of certain seed-borne and early season seedling diseases or pests, which without the use of Seed Treatment, would be very difficult to control.

The Seed Industry and Seed Treatment Applicators make an effort to reach an optimum application effect and consistency, as supported within a "Total Quality Management" (TQM) system. A TQM approach defines the treatment process, handling of treated seed, worker protective measures, protective measures for the environment, application techniques, operational techniques and information requirements to facilitate the safe and legal treatment of seed, handling of treated seed, and use and disposal of treated seed.

The TQM approach should also provide transparent seed quality and performance information that is addressed to users. It is generally understood that Seed Treatment requires adequate quality standards and necessary application methods. These should be agreed between the involved parties case-by-case and accepted as a balance between potential high standards and justified costs to achieve the standards.

SEED QUALITY FOR SEEDLING PRODUCTION

Seed quality has great impact on the quality of planting stock. For the last 20 years, the technology of producing seedlings has advanced greatly. Parallel to this advancement, seed quality also has improved dramatically.

Cone Collection

Careful attention to cone collection is critical to obtaining good quality forest-tree seed. Successful collection depends on understanding seed maturation and dispersal characteristics of each species, knowing local weather trends, and evaluat—ing crop quality, harvesting procedures, and cone-storage methods.

Seed Maturation

Cone collection should be gin only when seed is mature. Immature seed can bring about various problems including (1) slow and incomplete germination, (2) low-vigor seed, resulting in smaller seedlings, (3) greater susceptibility to disease, (4) reduced storage capability, and (5) increased incidence of abnormal seedlings. In addition, extraction of immature seed is more difficult than that of mature seed. Various maturation indicators, reflecting visual, physical, biochemical, or climatic changes, can be used effectively to prevent harvest of immature seed.

Cone color, bract color, seed wing color, scale color, and color and firmness of embryo and megagametophyte can be visual indicators of seed maturity. These indicators, though indirect and subjective, have proved reasonably practical in many instances. Cone moisture content, cone specific gravity, and embryo development can be physical indicators of seed maturity. Loss of cone and seed moisture is closely associated with seed ripening, and the decrease in cone moisture content and cone specific gravity has been used to indicate maturity. Of these two indicators, specific gravity (SG) is usually preferred because it can easily be determined in the field.

This method has been successfully applied to various pines (Pinus spp.) and true firs (Abies spp.) using flotation liquids such as water (SG = 1.0) and various mixtures of kerosene (SG = 0.80), light motor oil (SG = 0.88), and linseed oil (SG = 0.93). The ratio of embryo length to embryo cavity length, which can be determined quickly in the field with a sharp knife and a l0X magnifying hand lens, also can be used to judge maturity.

Changes occurring within conifer seeds can be biochemical indicators of seed maturity. On the basis of observed correlation of reducing sugar content and germination, Rediske recommended that Douglas-fir cone collection be initiated when reducing sugar content has fallen to 13 mg/g of seed weight. In a subsequent study, Rediske and Nicholson found that, in noble fir, the increase in crude fat content is more closely related to seed maturation and recommended the threshold value of 250 mg/g of seed for beginning cone collection. Although measuring biochemical indicators is time consuming and requires special laboratory equipment, it is thought to be more reliable than methods based on visual observation.

Changes in temperature, particularly during the summer in which seeds mature, can strongly influence the rate of seed maturation and are used as climatic indicators. Consequently, degree-day summations should be potentially more reliable than calendar date, especially at high latitude or high altitude, where summer temperature may limit seed development. Tanaka and Cameron reported that 1,310 degreedays are required for ponderosa pine seed to mature at high elevations in southeastern Oregon. Zasada related cone and seed development in white spruce [*Picea glauca* (Moench) *Voss*] to summer heat-sum and found that 625 degree- days were required to produce cones that could be successfully after-ripened in Alaska.

Seed Dispersal

Although seed-maturation characteristics have been exten—sively studied, little is known about timing of seed dispersal in relation to cone characteristics or climatic variables. Most observations relate the timing of seed dispersal to calendar dates, but such correlations may be of little value in the field because of yearly variation in weather patterns. Together with knowledge of seed maturation rate, approximate seed-dispersal dates could be of practical importance to cone collectors. The field observations made by our laboratory also indicated that the earlier seed matured, the more quickly it started disseminating, probably due to faster drying of cones. Similar observations should be of value in capturing the maximum seed yield of other conifers that have a responsive reflex of cone scales.

Artificial Ripening

It is important that cone collection be initiated after seed has attained full maturity. However, immature seed can be artificially ripened during cone storage in certain species. Artificial ripening has been successful on noble fir, grand fir [*Abies grandis* (Doug]. *ex D. Don*) *Lindl.]*, white fir, and Nordmann fir [*Abies nordmanniana* (Stev.) *Spach*]. Because of this potential increase in germination during cone storage, true fir cones are usually stored longer than those of other conifers before seed extraction. Douglas-fir, several species of pines, and white spruce have also shown increased germination during artificial ripening. However, despite the findings of various researchers and the potential benefits, artificial ripening has not been extensively used for conifers other than true firs in the Northwest-probably because there is more risk of poorer germination and reduced seed yield in other species.

Weather

Weather conditions significantly impact cone collection. Except for pines with serotinous cones or some cypresses (*Cupressus spp.*) or junipers (*Juniperus spp.*) for which year-round collection is possible, the optimum cone-collection period for most conifers at any given location is relatively short. This period occurs sometime between late summer and late fall but could vary by up to 2 to 3 weeks depending on weather conditions. For example, if the snow melts late at high elevations during a cool spring, flowering may be so late that seed maturation could be delayed significantly. A hot, dry summer may shorten the optimum cone-collection period by causing early seed fall, whereas cool, rainy conditions may delay it. Seed generally ripens earlier at lower elevations and on south and west slopes and later at higher elevations and on north and east slopes.

In addition to general spring and summer weather trends that determine seed-maturation and dispersal patterns, weather conditions during the cone-collection period itself are also important for the cone-harvesting operation. High winds or rain may preclude tree climbing, disrupt access to collection areas, and reduce pickers' productivity. In many areas in the Northwest, a drying east wind during the fall collection period may cause seed to disseminate too quickly, thereby reducing seed yield. For these reasons, daily forecasts and 5-day out -looks are valuable aids to coordinating cone-collection activities.

Crop Quality

Once seed maturity has been determined, the quality of cone crops to be harvested must be evaluated. This generally is done by estimating the number of good seeds present in several representative cones, sliced lengthwise with a sharp knife. Cones of Douglas-fir, western hemlock, and pines are sliced through the center; those of true firs are cut lengthwise ¼ to ½ inch to one side of center. A variety, of knife assemblies is available for slicing conifer cones. Minimum acceptable seed-count requirements may vary from year to year according to supply and demand. Average good seed counts are 6 for Douglas-fir, 8 for western hemlock, 10 for ponderosa and lodgepole pine, and more than 50% of the seed (if seed has good appearance) for noble and white fir.

Lodgepole pine in certain areas produces cones that are very hard and, therefore, difficult to section. To extract seeds, such cones can be dipped in boiling water for 10 seconds, then placed in an oven at 65°C for 3 to 4 hours. A minim um of 20 filled seeds per cone is required before a crop can be harvested. In addition to the filled-seed count, damage by biotic agents such as insects and disease, climatic extremes, or other abnormalities also should be assessed because these affect seed yield and are important factors in selecting areas from which to collect. Dobbs et al. do not recommend collection if more than 50% of seeds are damaged. Several articles may be of help in identifying and assessing insect and disease damage.

Collecting Methods

Cones are collected from western conifers: (1) by climbing standing trees, (2) from felled trees, and (3) from squirrel caches. Collecting cones from standing trees-the surest method to harvest seed of known origin, qua lity, and maturity-is often time consuming, expensive, and dangerous. Cones can be picked much more easily from felled trees in logged areas, but pickers should ascertain whether seeds were sufficiently mature when the trees were felled. Cones should be picked immediately after felling so as to minimize seed loss due to cone opening or mammal, bird, and insect damage. Squirrelcached cones are easy to collect, but their use is sometimes questioned because the source and quality of the crop tree are not known. No evidence suggests, however, that seeds collected by squirrels are inferior to those collected by other means. Helicopter collection has been experimentally tested in Canada by Dobbs et al.. Mechanical tree shakers, regularly used on southern pines, have been tried only experimentally for western conifers. Although not easily adaptable to Northwest terrains for natural-stand collection, mechanized cone collection should play an important future role when western seed orchards are in full production.

Cone Storage

Cones are stored (1) because processing equipment is not usually capable of extracting seeds from all harvested cones at once ; (2) to decrease cone moisture content, thereby reducing kiln drying time; and (3) to artificially ripen seeds of species such as true firs and improve seed-germination potential. In large-scale cone collection, cones are usually placed in burlap bags, which are stored either temporarily near collection sites or in storage sheds at the extractory.

However, great care should be exercised to maintain seed quality during cone storage. Burlap bags should not be filled to the tops, so that cone scales can fully expand upon drying; if scales cannot open sufficiently, seed extraction may be severely impaired. Burlap bags should not be stacked up in large piles; this can lead to seed losses due to overheating or to insect and disease damage. Warm, moist environments can harm seed quality ; hence, good ventilation should be provided. At a few seed-processing plants, cones of true firs and spruces are stored on ventilated mesh screens for artificial ripening.

There has been an attempt to rank the different species according to relative ability to withstand prolonged cone storage. The ranking is primarily based on intuition and experience, but such information is valuable in scheduling cone processing. Several studies conducted with western species have confirmed the success of current cone-storage practices and have shown that, if cones are handled properly and storage conditions are optimum, seed could be safely stored in intact cones for up to 4 to 6 months. For cone storage beyond 4 months, it may be advantageous to install frost protection because subfreezing temperatures could significantly reduce seed germinability. After cones are harvested and stored, seeds are extracted and prepared for either immediate sowing or storage. This series of operations, called seed processing, includes kiln drying, cone tumbling, scalping, dewinging, and cleaning and sorting.

Kiln Drying

Given good drying conditions, cones of most conifers open readily. Under natural storage conditions, however, cones may not be thoroughly and uniformly dried, especially when weather is humid and cool. Cones should therefore be kiln dried to facilitate extraction.

Kilns are of two types: rotating and progressive. In rotating kilns, a batch of cones is loaded into and dried within a drum where temperature and humidity are usually controlled. Such kilns, although not suitable for drying a large quantity of cones, can provide specific drying temperatures and relative humidities for small-lot processing. In progressive kilns, loaded trays are moved at certain time intervals to expose cones to increasingly warmer air as they dry. This type of kiln is more suitable for large-batch processing.

Kilns are generally operated at temperatures between 32 and 60°C. Although studies have shown that the biologically lethal temperature of most tree seed is around 66°C, the operational maximum temperature should not exceed 43°C. Because cones often have a high moisture content after storage, however, drying should be started at low temperatures that are progressively elevated. Drying cones with high moisture content immediately at high temperatures should be avoided because it could lead to case hardening and result in partial cone opening and incomplete seed extraction. However, the problem of case hardening can, to some extent, be overcome by moistening scales or soaking cones in water.

Air humidity is as important a factor as temperature. Low humidity is the key to more complete drying. For example, cones can be successfully dried at the relatively low temperature of 32°C if relative humidity is below 30%. Cones of most major conifers in the Northwest readily open upon drying. However, lodgepole pine cones from certain geographic areas are serotinous and require a short soak in hot water before kiln drying. Additional soaking cycles with water have been reported to increase seed extraction by 20 to 84%.

Cone Tumbling

In rotating kilns, cones are dried and tumbled simultaneously, and seeds fall out as cones open. Generally, loose seeds drop through perforations in the drum. Cones dried in progressive kilns are subjected to shaking action by tumblers to extract seeds from cones. A tumbler is a rectangular or round wiremesh container mounted horizontally on its long axis, which turns at a slow speed. Small quantities of cones may be tumbled in batches. In large-scale continuous operation, the tumbler axis is inclined so that rate of cone movement through the tumbler can be regulated.

Scalping

Seeds coming from the tumbler must be separated from a mixture of cone fragments, hardened pitch, foliage, dust, and other debris. This step, called scalping, is achieved by vibration, air movement, or screens, alone or in combination. The most commonly used equipment has several layers of vibrating screens of different-sized mesh. Coarse materials such as scales and twigs are retained on the uppermost screen and slide down to be collected in one bin, while fine particles are screened to be deposited in another bin: the seed is usually collected through an intermediate screen.

Dewinging

Once debris has been eliminated, wings must be removed from many conifer seeds. Although wings are often loosened during tumbling and scalping, dry or wet dewinging may also be required. Dry dewinging, a technique which employs a rubbing action to remove wings from dry seed, is generally used for Douglas-fir, pines, and true firs. Small lots can be dewinged in a cloth bag; lots of up to 5 kg are better handled in a Dybvig macerator ; and large lots are best dewinged with a brush-type dewinger, although auger-type dewingers have also been used successfully.

Because dry dewinging is the processing step that is most likely to cause seed damage, extra caution should be exercised to use proper equipment and to minimize unnecessary friction. In one study, for example, three cycles of brushdewinging seeds of subalpine fir [*Abies lasiocarpa* (Hook.) *Nutt.*] destroyed 50% of the originally viable seeds. Because dry dewinging can mechanically damage seed, many seed workers prefer wet dewinging, especially for pines and spruces. The principle of wet dewinging is that wings are more hygroscopic than seed and, upon wetting, are released cleanly.

The Kason Vibrator and a rotating cement mixer with a soft brush have been successfully used for wet dewinging. However, because seed absorbs moisture during wet dewinging, it must be redried sufficiently before storage. Germination tests verified that a 20- to 30-minute water soak, followed by wet dewinging and air drying for 16 hours at 26 to 30°C to 4 to 8% moisture content, did not adversely affect seed quality.

Cleaning and Sorting

Empty seed, partially filled seed, and other foreign particles are removed from good seed in the final cleaning. Scalpers and fanning mills are often used for species that have few scales, such as Douglas-fir and pines, but vibratory gravity tables are best for true firs. Pneumatic seed cleaners have also been successfully used for various conifer species. All this equipment, in combination, further improves sorting efficiency. Flotation sorting with water, alcohols, and other organic liquids has been used to clean red spruce (*Picea rubens Sarg.*), true firs, and several pines, although this method has only been tested experimentally with western species.

A noteworthy development in seed sorting is the IDS (incubation-drying-separation) method, developed by Simak, which can separate nonviable, as well as empty and partially filled, seed from viable seed. Fully imbibed seed is first incubated for a short time, then gradually dried, and finally separated by various specific- gravity methods. Because empty and nonviable seeds lose water more quickly during the drying phase, differences between nonviable and viable seeds are magnified, making subsequent separation by standard gravity methods more effective. Scots pine (Pinus sylvestris L.) seed of low germinability was successfully upgraded by this method experimentally.

In US, most processing work is done at seed plants in the Northwest. All eight seed-processing plants responding to the OSU Nursery Survey process their own seed as well as seed harvested by other organizations. However, four of the 16 nurseries responding do at least part of the processing at their own facilities. The remaining nurseries have private or state plants process their seeds.

Seven of the eight seed-processing plants and seven of the 16 nurseries set their own standards of purity for commercial seed transactions or nursery sowing (Table 1). The seed plants and nurseries replying to the Survey had a generally higher standard of purity than the Western Forest Tree Seed Council recommendations for four of the six major coniferous species; the lower accepted purity standards of the true firs (Table 1) may indicate possible difficulties in removing nonseed components without adversely affecting seed germination.

Table 1. Minimum purity standards recommended by the Western Forest Tree Seed Council

Species	Tree Seed Council (%)	Seed plants andnurseries (%)
Douglas-fir	95	95-99
Ponderosa pine	95	95-99
Lodgepole pine	90	99
Noble fir	95	90-98
White fir	95	90-98
Western hemlock	90	95

Seeds of true firs are known to be especially sensitive to handling and mechanical damage.

SEED TESTING

Seed testing evaluates seedlot quality and is essential for both seedling production and commercial seed transactions. Most tree-seed tests are conducted with methods based on rules of the Association of Official Seed Analysts (AOSA) or the International Seed Testing Association (ISTA).

Sampling

The first step in seed testing is to draw a sample that represents the entire seedlot. A seedlot is defined as a unit of seed of reasonably uniform quality from a particular location or elevation. Seedlot size varies with testing rules and among laboratories. ISTA, for example, has determined that a seedlot should be less than 5,000 kg for seeds the size of beech (Fagus spp.) seed or larger, or 1,000 kg for seeds smaller than beech. The Western Forest Tree Seed Council recommends that lots in excess of 227 kg be divided into equal smaller lots for sampling. Loose seeds in containers should be sampled with seedsampling probes long enough to reach all areas in the containers. The sample should be composed of equal portions taken from evenly distributed volumes of the lots to be sampled, each sample proportional to the size of the container.

Physical Characteristics

Purity

Purity tests measure the percent by weight of four major components: (1) pure seeds of the test species, (2) seeds of other crop species, (3) weed seeds, and (4) inert matter (leaves, cone scales, etc.). The purity test is usually the first test per—formed for a given lot and is especially important for commer—cial transactions, which are based on weight.

Moisture content

Seed moisture content is most often determined with the air-oven method. Seed samples are heated in ovens; the weight loss that occurs during drying is considered to be seed moisture. ISTA rules prescribe oven drying at 105°C for 16 hours for all tree seeds except those of the genera Abies, Cedrus, Fagus, Picea, Pinus, and Tsuga. Seeds of those genera contain a significant amount of volatile oils and resins which may be lost at the above temperature. Therefore, their moisture content must be determined by toluene distillation. Electronic moisture meters, though not as accurate as the above methods, are frequently used by various seed workers; they give rapid measurements desirable, for example, when checking moisture in a large number of seedlots being dried before storage. Seed moisture content can be expressed as a percentage of water loss of either total fresh weight or corresponding ovendry weight. Seed moisture content has been expressed on a dry-weight basis in some research, but international usage is exclusively on the fresh-weight basis.

Weight

Seed weight, required for calculating sowing rates in nursery sowing and direct seeding, is a function of seed size, moisture content, and proportion of full seed in a given lot. The commonly used unit is the weight of 1,000 pure seeds (1,000 seed weight). ISTA specifies weighing eight random samples of 100 seeds each from the pure-seed component; however, some laboratories use two or more samples of 500 seeds each. When means of replicates vary more than 10%, additional samples should be weighed. All weights should be accurate to three significant digits.

Biological Characteristics

Tests to estimate seed viability

Germination potential, perhaps the most important quality measurement in seed testing, is used to determine sowing rates as well as whether seed must be sown immediately or can be stored. Seeds of different species have different requirements for optimum germination. This potential can be (1) evaluated directly by germinating seeds under predetermined conditions or (2) estimated indirectly with biochemical staining, embryo excision, cutting tests, x-ray radiography, or hydrogen peroxide tests.

The most reliable method is germination in a controlled environment. At least 400 seeds, usually divided into four replicates of 100 seeds each, from the pure-seed component of the purity test are normally prechilled for up to 28 days and germinated on suitable substrates. Substrates should (1) be nontoxic, (2) be free of molds or other microorganisms, and (3) provide adequate aeration and moisture ; those recommended by AOSA are blotter papers, paper towels, washed sand, vermiculite, perlite, and peat moss.

Most (over 70%) of the coniferous species listed in the AOSA rules are germinated under alternating temperatures (30°C for 8 hours in the light, 20°C for 16 hours in the dark). An intensity of 750 to 1,500 (± 250) lux [75 to 150 (± 25) foot—candles] is recommended. Seed is counted as germinated when all essential structures appear normal. Retests are necessary when an extremely high proportion of full, ungerminated seed is left at the end of the test, or when variation among test replicates exceeds the accepted tolerances.

Although controlled-environment germination tests are reliable, they are often time consuming, especially for dormant species requiring prechilling. Several rapid methods of estimat—ing viability have been proposed, two of which-tetrazolium staining and embryo excision-are now recognized as official testing procedures.

The tetrazolium test is the most commonly practiced biochemical staining method. Seeds are immersed in 2,3,5-triphenyl tetrazolium chloride. Living cells stain red as tetrazolium is reduced by dehydrogenase enzymes to form a stable red triphenyl formazan, which is insoluble in water. The method is fast but lacks uniformity in staining ; therefore, results can be difficult to interpret. Other biochemical staining methods applied to seed testing with varying degrees of success include those using salts of selenium and tellurium and Indigo Carmine.

The excised embryo test is recommended for several species of pines including Coulter pine (*Pinus coulteri D.* Don), Jeffrey pine, and sugar pine (*Pinus lambertiana Dougl.*). Excised embryos are cultured on moist filter or blotter paper in covered dishes under light for 10 to 14 days at 18 to 20°C. Viable embryos remain firm and white and turn green, indicating growth, whereas dead ones turn dark or are covered with mold. This method is fast but requires skilled analysts.

Other quick methods include the cut, x-ray, and hydrogen peroxide tests. In the cut test, seed is bisected and then rated visually; this is the simplest but most unreliable method because distinguishing seeds damaged during handling and storage is very difficult. The x-ray test is fast, especially when Polaroid film is used, and development of contrast techniques has greatly expanded x-ray test capabilities. Disadvantages are difficulty in interpretation and relatively high equipment costs. The hydrogen peroxide (H_2O_2) test allows assessment of root growth in 1% H_2O_2. It is simpler to perform than t he excised embryo test and is more objective and easier to interpret than x-ray. However, as with other quick tests, it tends to overestimate viability, compared with germination tests.

Seed vigor

Nursery bed germination is usually slower and less complete than laboratory germination. Therefore, various laboratories have attempted to define and determine seed vigor to improve prediction of nursery germination. Three major groups of expressions have been proposed: (1) mathematical values based on standard laboratory test results, (2) germination under stressful conditions, and (3) biochemical testing. Mathematical expressions have been most widely tested.

They include the number of days required to attain a certain proportion of total germination, germination value, modified germination value, and the Weibull function. Germination under stressful conditions has been developed mainly for seeds of agricultural species; most widely used are the cold test for corn and the accelerated aging test for soybean. However, application of these tests or development of new procedures for tree seed has been rather limited. Biochemical tests have been tried to a limited extent; the few reported include tetrazolium staining and the GADA (glutamic acid decarboxylase activity) test.

SEED LONGEVITY

The life span of seeds varies with species. Seeds are classified into three biological categories according to their life span under natural conditions: (1) microbiotic seeds (life span not exceeding 3 years), (2) mesobiotic seeds (life span from 3 to 15 years), and (3) macrobiotic seeds (life span from 15 to more than 100 years). Seeds of most conifers and hardwoods are microbiotic. Under regulated storage conditions, however, longevity of many tree seeds can be extended more than tenfold. For example, the viability of naturally dispersed seed of spruce and many pines extends only into the first growing season and, occasionally, into the second growing season.

Under subfreezing storage, seed viability of these same species can easily be maintained at high levels for 10 years or longer. Storage over 10 years is not usually required for seedling-production purposes but may become vital to future tree-breeding programs. Under optimum storage conditions, seed viability of certain trees might be maintained indefinitely, but the maximum potential for maintaining original seed viability has not yet been determined for most species.

Seed quality has a significant impact on storage capability. Factors affecting quality are seed maturity, cone handling, and seed extraction and processing. Immature seeds are not only poor in germinability and liable to be further damaged by seed processing but also are difficult to store successfully. Overheating during extraction and damage caused by dewinging also have been found to adversely affect seed quality. Injured seeds are not suitable even for short-term storage because they have a high rate of respiration, undergo spontaneous heating, and deteriorate rather quickly.

Of all the factors influencing seed storage, moisture content may be the single most important one in maintaining germinability. Various researchers have demonstrated the detrimental effect of high seed moisture on tree-seed viability; increased rates of respiration and changes in carbohydrates and fats presumably cause seeds to use their food reserves. Excessively low seed-moisture content also may reduce storage capability. Some species, including Douglas-fir, can tolerate drying to 0% moisture content ; however, overdrying can destroy the monomolecular layers that protect against oxidation. Recommended seed-moisture content for storing Douglas-fir, ponderosa pine, lodgepole pine, and western hemlock is 6 to 9% (wet-weight basis); that for true firs is 9 to 12%. However, Danielson and Grabe showed that optimum moisture content for noble fir is also 6 to 9%.

The effect of storage temperature on the retention of treeseed viability has been thoroughly investigated. The general relationship between storage temperature and moisture content was described by Barton as follows: at a given moisture content, the higher the storage temperature, the faster the deterioration of seed viability: the lower the storage temperature, the greater the tolerance to high moisture content and the better the retention of viability. Some studies have shown that temperature slightly above or below freezing may be sufficient to prevent deterioration for short -term storage, but the retention of viability has generally been better when seeds were stored at -18°C, particularly for longer periods.

Tree seeds can be stored either wet or dry. Large seeds of hardwood species require moist conditions and are usually kept wet for short-term storage, whereas small seeds of conifer species, including all major conifers in the Northwest, are stored dry. Seed moisture content is controlled by storing properly dried seed in tightly closed containers or by regulating humidity in the storage area (as for many agricultural seeds); in the Northwest, dried seed is generally placed in closed containers, although some facilities do use humidityregulated storage rooms. These facilities, although costly, are effective in minimizing reabsorption of moisture by dried seed, especially in areas with humid climates.

The most frequently used storage containers are plastic bottles with screw tops, polyethylene bags, and fiberboard drums. More than 80% of the organizations surveyed store their seeds in rigid drums lined with polyethylene bags. This method is common in many areas because it is relatively inexpensive and effectively prevents uptake or loss of moisture by seed from the atmosphere. Of 16 organizations, nine use plastic bags as containers; five of those use I- to 6-mil plastic, and the remaining four use 7- to 8-mil plastic. Thin bags are subject to ripping but easier to handle when cold. However, plastic containers are not completely impermeable to moisture. Use of thicker materials may be desirable for seed requiring low seed-moisture content if external humidity is high and seeds are to be stored for a long period.

RETESTING

Retesting is often recommended for seeds stored for a relatively long period (5 years or more). Even under ideal storage conditions, certain poor -quality seedlots rapidly lose their viability. Fourteen of 16 organizations conduct viability tests at 2- to 6-year intervals. Seed moisture content is retested by only three of the 16, however, probably indicating that moisture content does not fluctuate significantly under current storage conditions.

Even if seed moisture changes only minimally in storage, additional moisture could be introduced when seeds are withdrawn. Therefore, it is particularly important that sealed containers removed from cold storage be permitted to reach ambient temperatures before being opened to avoid condensation of water within the container. Sahlen and Bergsten found that temperature was completely equalized in the center of a 28-liter container, with walls 2.5 mm thick, 36 hours after the container was moved from a -16°C storage room to 22°C ambient temperature. To minimize repeated opening and resealing and to reduce storage space, the use of smallsized containers (10- to 25-kg capacity) has been recommended.

SEED STRATIFICATION

Tree seeds. unlike agricultural seeds, are in many cases characterized by deep dormancy. This is true for most Northwest conifers. Seeds of different species or different geographical origins often require different pretreatments and conditions for optimum germination. The most commonly used pretreat—ment to break dormancy is stratification-which usually is moist cold treatment for up to several months. Stratification is generally known to bring about changes in anatomy or physiology, including embryo growth, and in metabolism. Physiologically, breaking of dormancy has often been explained in terms of a shift in the inhibitor-stimulator balance. Presumably, although it may not directly affect the level of inhibitors, stratification could increase growth-stimulator levels, which would then counteract the effects of inhibitors in breaking dormancy.

Successful cold stratification requires: (1) proper moisture content, (2) low temperature, (3) adequate aeration, and (4) proper length of time. In practice, seed originally was stratified by placing it between moisture-holding media such as peat moss or sand in boxes, tanks, trays, and oth er suitable containers and maintaining it there under cold, moist conditions. Some nurseries use outdoor soil pits. More recently, stratification in polyethylene bags has become common at many nurseries and seed-processing plants. This method, called "naked stratification," requires no moisture-holding medium and less effort in preparing seed for subsequent sowing. Seed is soaked in water in containers lined with plastic or mesh bags, drained of excess water, and kept at low temperatures for a predetermined period of time; bags are often loosely fastened to allow aeration.

Water Soaking

The rate of water absorption varies among species. Most conifers require 1 to 2 days of soaking to achieve full imbibition. It has been suggested that warm water can speed up water absorption by seed, and that running water and aeration can improve oxygen availability; however, this has not yet been substantiated experimentally for Northwest conifers. During soaking, four organizations aerate water, whereas three use running water. These practices are probably beneficial, although the effectiveness should be determined for each species.

Temperature

After draining, seeds are stored in the fully imbibed state. A few species, such as yew (Taxus spp.) and yellow-cedar [Chamaecyparis nootkatensis (D. Don) Spach], require storage at warm temperatures before cold storage: however, most coniferous species require low temperatures throughout. For loblolly pine (*Pinus taeda L.*), McLemore found optimum stratification temperature to be 10°C, but Robinson et al. reported that, for this same species, gradually increasing temperature over a 4-week period gave the best stratification results.

Temperatures above 5°C are not desirable because they increase the risk of overheating and subsequent deterioration, alt hough freezing temperatures also can damage seeds at high moisture content. Consequently, low temperatures of 2 to 5°C have been adopted as accepted operational practice in most cases. All except one organization use stratification temperatures between 1 and 5°C; the exception uses 0°C. Even in this case, however, embryos would not experience freezing due to their osmotic potential, which is lower than that of water. Premature germination during prolonged stratification can be minimized if seeds are held at 2°C, rather than 5°C, for both Douglas-fir and ponderosa pine.

Aeration

Aeration during stratification is necessary to supply oxygen for seed respiration and to allow carbon dioxide and heat to escape. Lack of aeration could therefore lead to deterioration of seed quality through buildup of toxic substances. The most commonly used technique is to leave a small air space at the neck of each bag in which seed is stratified and to massage the whole bag periodically. A few Northwest nurseries also use fine-meshed bags hung so that air may circulate.

Duration

Optimum stratification length varies among species and seedlots. In general, the longer the stratification period, the greater the rate of germination, especially under suboptimal germination temperatures. For this reason, seed destined for colder environments must be stratified long enough for quick and complete germination. However, prolonged stratification can cause seeds of some species to germinate prematurely; furthermore, vigor and total germination may be reduced if seeds are stratified for excessively long periods.

Premature germination sometimes occurs during prolonged stratification. Although premature germination is a serious concern in nurseries because the fragile seeds can be damaged in handling or during mechanical sowing, redrying and storing of stratified seed are possible. Danielson and Tanaka reported that ponderosa pine seed air-dried to 26% and Douglas-fir seed air-dried to 37% can be stored for 9 and 3 months respectively without losing the beneficial effect of stratification or having their viability adversely affected. Subsequently, Edwards tested the efficacy of surface-drying true fir seed after 1 month of stratification at saturation moisture, followed by 3 months of storage at 35% moisture content; this treatment not only prevented premature germination but also improved total germination and germination rate.

The exact mechanism behind the benefit of surface drying is not completely understood. It may be related to improved gaseous exchange brought about by removing the water film from the seed surface, which increases oxygen availability to the seed and facilitates the release of any accumulated toxic gases. Although the surface-drying technique provides the option of storing stratified seed for prolonged periods without losing stratification effects, the lower limit of seed moisture content should be determined for each species. Seeds can be stored safely below certain thresholds but seem to then require restratification after storage. This induction of secondary dormancy suggests that seed moisture and dormancy are closely related.

Other Treatments

Although stratification is an effective method to break dormancy, it is often time consuming. Past research has shown that hydrogen peroxide, gibberellic acid, ethylene, microwave irradiation, or osmotic agents can stimulate germination of conifer seed. However, these studies were usually conducted under optimum germination conditions and may not be effective under the suboptimal temperature conditions frequently encountered in the field in early spring when seed is sown at Northwest bareroot nurseries. Further work is required to develop a quick, effective method that would facilitate germination under a wide range of temperatures and that would either eliminate the need for stratification or shorten the stratification requirement.

MEASURES OF SEEDLING QUALITY

The carbohydrates are a diverse group of compounds sharing common building blocks-the sugars. Within this group are hundreds of compounds and, in the case of the sugars. those compounds are rapidly interconverted. This complexity can be reduced, however, by grouping the carbohydrates according to their functions in the plant.

Structural Carbohydrates

In the seedling's carbon budget. the structural carbohydrates represent a non-refundable investment of carbon. Carbon spent by the plant to build structure is thereafter unavailable for other uses.Cellulose is the predominant structural carbohydrate in trees. It is a linear chain (or polymer) of glucose molecules that cross-link with adjacent cellulose molecules. This cross-linkage gives the plant much of its strength. Cellulose. like other structural carbohydrates, is found almost exclusively outside the cell membrane in the cell walls.

The hemicelluloses are a second important group of structural carbohydrates occurring in the cell walls. They include complicated polymers consisting of a wide variety of sugars. There has been some speculation that hemicelluloses may at times serve as reserve materials in trees, but the evidence is inconclusive. Their exact function in the cell wall is not yet known. Finally. the pectins. which are chains of modified and unmodified galacturonic acid molecules, appear in the middle lamellae between cells. They serve as an intercellular glue and are apparently not used as reserve substances.

Reserve Carbohydrate

Starch is the most important reserve carbohydrate. It comprises the "savings account" from which a seedling can draw if respiratory or growth expenses exceed photosynthetic income.Like cellulose. starch is composed of joined glucose molecules. Unlike cellulose, the starch molecule is easily disassembled by enzymes, a process that releases its constituent sugars. Starch occurs within special structures- amyloplasts and chloroplasts- inside the cells. In a rice grain or potato tuber. starch concentrations may reach 70% of the dry weight. In conifer needles and fine roots. starch concentrations may reach 30% of the dry weight. On the other hand. starch concentrations may drop to near zero in a root or leaf about to be shed. The sugars are the metabolically active carbohydrates and are involved in all biosynthetic and maintenance activities. They are the only form of carbohydrate transported from place the place within the plant. They can be regarded as the seedling's "pocket money." The sugars also serve as the major carbohydrate for storage when temperatures are low. Any starch present is likely to be converted to sugar during cold-hardening because the sugar acts as an antifreeze within the cells.

Carbon Inputs and Outputs

Photosynthesis

Photosynthesis is the process whereby CO_2 is drawn from the atmosphere and converted to simple sugars. The energy source for this process is sunlight. Photosynthesis represents the plant's income, its paycheck. All other chemical reactions within the plant depend upon the "burning" of these sugars to release energy. This burning of sugars, releasing CO_2, is what we mean by respiration. Respiration can be separated into that required for construction of tissue and that associated with maintenance of tissue.

Respiration Growth

The seedling has two costs associated with growing new tissue. The greatest cost is that of the carbon actually being placed into the new tissue-for example, the sugars that are placed into cellulose polymers. The second cost is for the sugars burned to provide energy for synthesis of new molecules. In the case of cellulose, this is simply a matter of joining sugars. To make more complicated compounds, however, requires considerably more energy. The CO_2 released in these processes is attributed to "growth respiration."

The rate of growth respiration depends upon the growth rate and the biochemical composition of the tissue. Note that by this definition, growth is an irreversible commitment of resources. Because it is not irreversible, an increase in weight as a result of starch accumulation is not growth: the starch can be burned and the weight lost.

Respiration Maintenance

Respiration rates never fall to zero in living tissue; there is a cost associated with tissue maintenance. This cost is usually termed "maintenance respiration." It is associated with the burning of sugars to yield the energy needed for maintenance processes. These processes include replacement and repair of enzymes and cell membranes as well as transfer of materials from cell to cell. Because enzymes and membranes are replaced more fre-quently at higher temperatures, maintenance respiration increases exponentially with increasing temperature, approximately doubling with every 10°C increase.

Accumulation and use of Reserves

The seedling's only source of sugars, its only income, is through photosynthesis. The seedling does not necessarily use its sugars immediately, however. In fact, a large proportion of the sugar is often stored as reserves. The exact amount being stored depends in large measure upon the rates of photosynthesis and respiration. Therefore, reserve accumulation is favored by anything that (1) increases photosynthesis, (2) decreases growth, or (3) decreases rates of maintenance respiration. Alternatively, any of the opposite effects will reduce reserve accumulation and eventually begin to deplete reserves.

Light

Within a limited range, photosynthesis is sharply increased by increasing irradiance. If such an increase does not cause an equivalent increase in growth, starch will accumulate. Alternatively. placing a seedling in darkness causes it to deplete its reserves, especially if growth continues. Only by using reserves accumulated during the day can seedlings continue to grow at night. High seedbed densities necessarily reduce the amount of light available to each seedling and may reduce starch and sugar reserves.

Plant Moisture

As plant moisture stress develops, growth is reduced long before there is any reduction in photosynthesis. Consequently, starch concentrations increase sharply under mild moisture stress. Of course, once a drought becomes severe, photosynthesis is also halted. Accumulation of starch under mild water stress provides a means by which a plant may better survive a period in which photosynthesis has been halted.

Nutrition

In some species, mild nutrient deficiencies also reduce growth more than they reduce photosynthesis. New growth requires major nutrient inputs, whereas photosynthesis can continue with nutrients already in place. For these reasons. a decline in nutrient availability will result in starch accumulation. Improved nutrient availability, on the other hand. may increase growth rate and reduce accumulation of reserves.

Seasonal Growth Patterns

The periods of growth of many tree species are controlled by daylength or chilling requirements rather than by resource limitations. Analysis of the processes discussed earlier nonetheless applies; starch and sugar accumulate before and after periods of growth for as long as photosynthesis is active. This accumulation explains the increases in starch concentrations commonly observed in the spring and late summer and the increase in sugar concentrations commonly observed in autumn.

Determining Carbohydrate Content Procedures

The carbohydrates comprise a complex group of compounds. To a certain extent, they can be analysed according to these functional groups as well.

Iodine Staining

Iodine staining of starch in intact tissue has been suggested as a measure of seedling quality ever since the subject of physiological grades was first introduced. Hellmers showed almost complete depletion of starch in shoots kept in cold storage for 4-1/2 months. Given its simplicity and low cost, this technique should be extremely useful as long as starch is the compound of interest.

Unfortunately, sugars do not stain with iodine: consequently, it is difficult to know whether a decrease in starch concentration means that starch is being respired or being converted to sugars. Nonetheless, iodine may well provide the most useful qualitative measure of carbohydrate status in the field. Iodine staining would be of little use, however, during the winter, when starch is converted to sugars. Because conversion to sugars occurs to a lesser extent in roots than in shoots, iodine staining might be useful even during the winter if applied to root tissues.

Extractions

Determination of the amounts of starch and sugar in a tissue requires that the starch and sugars be completely extracted. Because such extraction has proven difficult in the past, estimates of starch and sugar concentrations have varied widely. Sugars can be extracted with ethanol, methanol, or methanol:chloroform:water (MCW). Hot 80% ethanol has been most usual. The alcohol extract contains phenols, organic acids, and pigments as well as the sugars. It is necessary to remove these contaminants to determine the sugar concentrations colorimetrically. The non polar pigments (including chlorophyll) and the phenols can be partially removed by a 100% acetone extraction prior to the ethanol extraction. The remaining non polar contaminants can then be removed by adsorbing them out of the ethanol extract with insoluble polyvinylpyrollidone. Organic acids can then be removed by precipitation with neutral lead acetate; a clear, nearly colorless extract is left.

Solvent fractionation (addition of chloroform to MCW extract) may also be used to rid the extract of non-polar contaminants. After the soluble sugars have been removed. starch can be extracted from the residue. Enzymes are often used in this extraction because of their specificity. Unfortunately, the starch-extraction enzymes are easily deactivated by any phenols remaining in the tissue or in solution and therefore may not yield complete extractions. This problem is apparently much more severe in conifer tissues than in the crop plants for which most of these methods were developed. Haissig and Dickson may have circumvented this problem by extracting with MCW prior to starch extraction. but their method has not been rigorously checked for completeness of extraction.

Non-specific Carbohydrate Determinations

Total hexose sugar concentrations of sugar extracts and starch extracts can be determined with a spectrophotometer by measuring color developed in the reaction with anthrone. The major hexose sugars appear to yield very similar color development per glucose equivalent.

Specific Carbohydrate Determinations

Finally, if one wants to identify specific sugars in the extracts, it is necessary to use chromatographic techniques. Alternatively. glucose concentrations can be determined from color development after adding glucose oxidase to the solutions. Given the ready interconversions among sugars and the utility of the anthrone reaction, however. there seems little point in trying to identify specific sugars in assessing seedling quality.

Predictive Abilities

Correlations between survival and carbohydrate status have been demonstrated for many species, but this relationship has not been found consistently. There are a number of problems in interpreting the literature. First, the diverse carbohydrate extractions and the analyses used in these studies vary in completeness of extraction and in what carbohydrates they extract. Second. differences in growth conditions, handling, storage, and planting conditions could easily induce or alleviate carbohydrate stress in the seedlings. Just as one would not expect a fertiliser response if the added nutrients were not limiting growth, so one should not expect a carbohydrate response if carbohydrate status were adequate. In the following section discuss, how carbohydrate status affects seedling growth and survival.

Potentials of Root Growth

It has long been recognised that root growth is essential to the survival of the planted seedling, and a specialised literature has developed around the measurement of "root growth potential" (RGP) of seedlings. It was long thought that the major effect of improved carbohydrate status was an increase in RGP: however. Ritchie has recently presented strong evidence against this hypothesis. He measured a steady decline in "non-structural carbohydrates" (presumably starch and sugars) in seedlings in cold storage. although RGP rose sharply and then fell. He concluded that changes in RGP were not directly correlated with concomitant changes in concentrations of reserve carbohydrates. Many experiments involving girdling, shading, and defoliation suggest that root growth, at least in conifers, is dependent upon current photosynthate rather than reserves.

Respiratory Costs

If carbohydrate reserves are not related to RGP, then how else could they improve seedling growth and survival? Carbohydrate status presumably measures different aspects of seedling quality than does RGP. RGP, being closely related to bud dormancy, seems to be a measure of readiness to grow. Carbohydrate status, on the other hand, may be more a measure of capacity to maintain tissue function under stress, i.e., to withstand periods when respiratory losses exceed photosynthetic income. If so. then the effects of carbohydrate status on growth and survival might appear only when seedlings are forced to draw heavily upon their reserves.

It should be possible to calculate the point at which the carbohydrate balance "goes into the red" by using a process much like balancing a checkbook. Starting with a measured starch and sugar concentration, respiratory expenditures could be subtracted and photosynthetic deposits could be added as the seedling goes through its lifting, storage, and planting cycle. Moreover, it should be possible to work in the opposite direction, to calculate the total respiratory costs a seedling is likely to incur, to define that total as a threshold concentration and then attempt to grow seedlings with carbohydrate reserves in excess of the threshold. Unfortunately, few measurements of the magnitude of these respiratory costs are currently available. Nonetheless.

A seedling enters the winter months having stored a certain amount of carbohydrate. This amount varies with growing conditions. One can assume that the planted seedling must wait about 3 weeks to achieve a positive carbon balance. This delay occurs because the soil is cold, a state that limits water uptake. and because the photosynthetic apparatus has been disorganised by storage and by any frosts that may occur. During this time, the seedling must continue to meet its maintenance respiration costs, an activity that brings the reserve carbohydrate concentration down to 13%. In addition, the seedling must invest a large amount of photosynthate in preparing leaves for photosynthesis and buds for shoot elongation. This investment may require as much as 5 mg of carbohydrate per day for perhaps a week.

Long-term Effects

Carbon debts seem to carry over from one developmental phase to the next and from one growing season to the next. Thus, Engelmann spruce seedlings planted with poor carbohydrate status suffered high mortality the following winter because they were unable to develop sufficient cold-hardiness. In a Swedish study, Scots pine seedlings outplanted late in summer with reduced carbohydrate status still had significantly reduced growth the third growing season after planting.It is also reasonable to expect that drought resistance will be reduced as a result of low carbohydrate status. Ritchie suggested that such reductions might occur because of an inability to adjust osmotically and retain positive cell turgor at low xylem water potentials.

REFERENCES

Devitt, B. 2020. Fertilisation of two improved seed production areas. *For. Res. Rev. British Columbian For. Service,* p. 52-53.

Georgia Crop Improvement Association, 2018. *Certification standards for forest tree seed (1961 crop, revised).* 5 p.

International Crop Improvement Association. 2019. *Minimum seed certification standards.* Publ. No. 19. International Crop Improvement Association. 111 p.

Bibliography

Agrawal, V.K. & Gupta, M.C. 2019. Loose smut of wheat - A treat to seed production. *Seed Res.*, 17(1): 55-68.

Allard, R. W. 2020. *Principles of Plant Breeding*. John Wiley and Sons, Inc. New York

Allen, G. S. 2020. A method of distinguishing coastal from interior Douglas fir seed. *Brit. Columbia Lumberman,* 44 (8): 26-30.

Almekinders, C. and N. Louwaars. 2019. *Farmers' Seed Production.* Intermediate Technology Publications. Pages 219-221.

AVRDC. 2020. *Vegetable production training manual.* Asian Vegetable Research and Development Center, Shanhua, Tainan, Taiwan.

Barnett, J.P. 2021. "Flotation in ethanol reduced storability of Southern pine seeds". *Forest Science,* 17, 50-54.

Baskin, C.C. and Baskin, J.M. (2018) *Seeds: Ecology, biogeography, and evolution of dormancy and germination.*San Diego, Academic Press.

Beck, D.L. 2024. "Hybrid corn seed production". In C.W. Smith, J. Betran, and E.C.A. Runge (eds.), *Corn: Origin, History, Technology, and Production.* Hoboken, New Jersey: John Wiley & Sons, Inc.

Bewley, J. Derek, and Michael Black. 2014. *Seeds physiology of development and germination. The language of science.* New York: Plenum Press. page 230.

Boyd, A.H., Dougherty, G.M., Matthews, R.K. & Rushing, K.W. 2015. Seed drying and processing. *In* W.P. Feistritzer, ed. *Cereal seed technology: a manual cereal seed production, quality control, and distribution.* Rome, FAO.

Chadha, M. L. 2013. *Improvement of brinjal. Advances in Horticulture* (Eds.: Chadha, K. L. and Kalloo, G.) Vol. 5 - Vegetable crops. 105-135.

Chaisurisri, K., Ponoy, B., Wasuwanich, P. 2016. Storage of *Azadirachta indica* A. Juss. seeds. The Embryon, Vol.2, No. 1, 19-27.

Chin, H. F. 2018. "Storage and Vigour". *Seed Science and Technology* 16:1-4.

Chopra, K.R., R. Chopra, G. Rabbani, and K.K. Thimaiah. 2019. "Seed production". Pp. 445-477 in *Pearl Millet Breeding.* New Delhi, India: Oxford and IBH Publishing Company Pvt. Ltd.

Clements, D.J. 2017. Wheat seed storage under tropical conditions. *In* A.R. Klatt, ed. *Wheat production constraints tropical environments,* p. 360-365. Mexico, DF, CIMMYT.

Del Rio, L. and C. Bradley. 2022. *Anthracnose of Dry Beans.* Department of

Plant Pathology, North Dakota State University. Fargo, ND. Extension publication PP-1233.

Devitt, B. 2020. Fertilisation of two improved seed production areas. *For. Res. Rev. British Columbian For. Service,* p. 52-53.

Diekmann, M. 2021. Seed-borne diseases seed production. Aleppo, Syria, ICARDA.

Diojode, S.D. 2021. *Seed Storage of Horticultural Crops.* Food Products Press. Binghamton, NY. Pages 235-240.

Doran, J.C., Turnbull, J.W., Boland, D.J, Gunn, B.V. 2013. *Handbook on seeds of dry zone acacias. Guide for collecting, extracting, cleaning and storing the seed and for treatment to promote germination of dry-zone acacias.* FAO, Rome

Douglas, J. E. ed. 2021. *Successful seed programs: A planning and Management Guide.* International Agricultural Development Service. Westview Press Boulder, Colorado.

Ehremberg, C., Gustafsson, A., Plym Forshell, C. P. & Simak, M. 2019. Seed quality and the principles of forest genetics. *Hereditas,* 41: 291-366.

Ellis, R.H. 2018. The viability equation, seed viability nomographs, and practical advice on seed storage. *Seed Sci. Tech.,* 16: 29-50.

FAO. 2016. *The world sorghum and millet economics-facts, trends and outlook.* Patancheru 502 324, Andra Pradesh, India: International Crops Research Institute for the Semi-Arid Tropics; and Rome, Italy: FAO.

Faulkner, R. 2022. Seed stands in Britain and their management. *Quart. J For.,* 56 (1): 1-15.

For. Com. Tasmania 2014. *Seed manual. Procedures for seed collection, handling, storage, purchase and accounting.* Forestry Commission, Hobart, Tasmania.

George, R.A.T. 2015. *Vegetable seed production.* Longman, New York. P223-229.

Georgia Crop Improvement Association, 2018. *Certification standards for forest tree seed (1961 crop, revised).* 5 p.

Gill, H. and Tomar, B. 2021. *Vegetable Statistics at Glance.* Indian Agricultural Research Institute, New Delhi, India. Tech. Bull. No. 4.

Granhof, J. 2014. *Seed extraction 2. Extraction of pine seeds by means of sun drying on elevated trays followed by tumbling.* DFSC Technical Note No. 17. Danida Forest Seed Centre. Denmark.

Gutterman, Y. (2022) *Seed germination in desert plants.* Springer Verlag, Berlin/Heidelberg.

Hagner, S. & Bergman, F. 2021. On insamling av tallkott från stående träd. [Collection of pine cones from standing trees]. *Norrlands Skogsv. Förb. Tidskr.,* 1: 15-391.

He, G.I., X. Zhu and J.C. Flinn. 2019. Hybrid seed production in Jiangsu province, China. *Oryza* 24: 285-296.

Heemskerk, W. 2021. *Seed selection and seed storage.* Ministry of Agriculture.

Hsieh, F., H.C. Huang, H.H. Mundel, et al. 2015. Resistance of Common Bean (*Phaseolus vulgaris*) to Bacterial Wilt Caused by *Curtobacterium flaccumfaciens* pv. *flaccumfaciens*. Journal of Phytopathology. 153 (4): 245.

International Crop Improvement Association. 2019. *Minimum seed certification standards*. Publ. No. 19. International Crop Improvement Association. 111 p.

International Workshop on Seeds, and G. Nicolas. 2021. *The biology of seeds recent research advances : proceedings of the Seventh International Workshop on Seeds, Salamanca, Spain 2002*. Wallingford, Oxon, UK: CABI Pub. Page 113.

IRRI. 2018. Rice Seed Health. Proceedings of the International Workshop on Rice Seed Health, 16-20 March 1987.

Janaiah, A. and M. Hossain. 2020. Hybrid rice for food security in the tropics: An evaluation of farm-level experiences in India. Paper presented at the 3rd International Crop Science Congress held in Hamburg, Germany, 17-22 August 2020.

Johnson, Jr., H. 2015. *Eggplant. Cooperative extension leaflet 21400*, University of California, Berkeley.

Kabay, D. and Lewis, A. 2019. "Collection, handling and storage of Australian native plant seed". In: *Germination of Australian native plants seed*. (Langkamp, P.J., ed.). Inkata Press. pp. 20-30.

Koike, S.T., P. Gladders, and A.O. Paulus. 2007. *Vegetable Diseases*. A Color Handbook. Academic Press, San Diego, CA.

Laverack, G.K. 2014. Management of breeders seed production. *Seed Sci. Tech.*, 22: 551-563.

Mamluk, O.F. & van Leur, J. 2016. Transmission of seed-borne diseases of wheat and barley. *In* J.P. Srivastava & T.L. Simarski, eds. *Seed production technology*. Aleppo, Syria, ICARDA.

Mao, C.S. 2018. Hybrid rice seed production in China. In: *Seed Health*. International Rice Research Institute, Los Banos, Laguna, Philippines. Pp. 277-282.

Mao, C.X. 2021. Improving seed production to speed up the global commercialisation by hybrid rice. 2001. In: *Rice Research* for Food Security and Poverty Alleviation, S. Peng and B. Hardy (Eds.). Proceedings of the International Rice Research Conference, 31 March-3 April 2020. International Rice Research Institute, Los Banos, Laguna, Philippines. 692 pp.

Maruyama, K. and K. Oono. 2021. Induction of mutation in tissue culture and its use for plant breeding. VII. On the female and male sterile rice plants regenerated from seed acalluses. *Jan. Jour. Breed.* 33 (Suppl. 1): 24-25. (In Japanese).

McCollum, J. P. 2018. *Producing vegetable crops*. Interstate, Danville, Illinois. 558pp.

McIntosh, Charles. 2018.*"The Practical Gardener, and Modern Horticulturist."* Thomas Kelly: London. Page 288.

Mew, T.W. and Misra, J.K. 2014. *A manual of rice seed health testing.*

International Rice research Institute. Los Banos, Laguna, Philippines. P.O. Box 933, 1099 Manila, Philippines.

Napier, I.A. and Robbins, A.M.J. 2018. *Seed storage in the nursery* (*Prunus, Alnus* and *Pinus* spp.). Banko Janakari, Vol. 1: 1, 13-20.

Nath, B. 2015. Sorghum (*Sorghum bicolour* L). Pages 165-171 in Zambia seed technology handbook. Ministry of Agriculture, Food and Fisheries.

Paged. 2022. *Forest tree seeds and trees.* Warsaw, Paged Foreign Trade Enterprise Export and Import.

Partridge, J.E. 2023. *Common Blight of Bean.* University of Nebraska-Lincoln. Lincoln, NE.

Schwartz, H.F., J.R. Steadman, R. Hall, R.L. Forster. 2015. *Compendium of Bean Diseases,* 2nd Edition. APS Press. St. Paul, MN.

Sherf, Arden F. and Alan A. McNab. 2016. *Vegetable Diseases and Their Control*: Second Edition. Wiley and Sons. New York, NY.

Singh, P. 2015. Pearl millet (*Pennisetum americanum* L.) pages 171-182 in Zambia seed technology handbook. Ministry of Agriculture, Food and Fisheries.

Som, M.G. and Malty, T.K. 2016. *Brinjal. In: Vegetable crops in India.* Bose, T.K. and Som,M.G. (eds.) Naya Prokash, Calcutta, India. P293-342.

Staff. 2006. *Pacific Northwest Plant Disease Management Handbook, Radish - White Rust.* Oregon State University Extension Service.

Stubsgaard, F. and Moestrup, S. 2021. *Seed Processing.* Lecture Note No. C-7. Danida Forest Seed Centre, Denmark.

The Maize Program. 2019. Development, Maintenance, and Seed Multiplication of Open-Pollinated Maize Varieties, 2nd edition. Mexico, D.F.: CIMMYT.

Thompson, H. C. and Kelly, W. C. 2017. *Vegetable crops.* McGraw - Hill, New York.611pp.

Thomson, J.R. 2019. *An introduction to seed technology.* Leonard Hill.

Tompsett, P.B. 2022. "The effect of desiccation on the longevity of *Araucaria hunsteinii* seed". *Annals of Botany* 50, 693-704.

Veira, C.P., Veira, R.D. and Paschoalick, J.H.N. 2014. "Effects of mechanical damage during soybean seed processing on physiological seed quality and storage potential". *Seed Sci. and Technol.* 22: 3, 581-589.

Wilcoxson, R.D. & Saari, E.E., eds. 2016. *Bunt and smut diseases of wheat:* concepts and methods of disease management. Mexico, DF, CIMMYT.

Worland, A.J. & Law, C.N. 2015. Aneupolidy in semidwarf wheat varieties. *Euphytica,* 34: 317-327.

Xu, S.J. and B. Li. 2019. Managing hybrid rice seed production. In: *Hybrid Rice.* International Rice Research Institute, Los Banos, Laguna, Philippines, pp. 157-163.

Yuan, L.P. 2015. A Concise Course in *Hybrid Rice.* Hunan (China): Hunan Science and Technology Press. 168 pp.

Index